Vahid Pirouzfar, Mastaneh Narimani, Ahmad Fayyazbakhsh, Chia-H
Milad Gharebaghi
Gasoline Additives

Also of interest

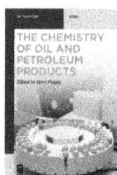
The Chemistry of Oil and Petroleum Products
Merv Fingas, 2022
ISBN 978311069436, e-ISBN 9783110694529

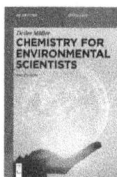
Chemistry for Environmental Scientists
Detlev Möller, 2022
ISBN 9783110735147, e-ISBN 9783110735178

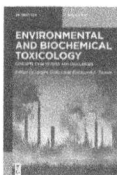
Environmental and Biochemical Toxicology
Concepts, Case Studies and Challenges
Jürgen Gailer and Raymond J. Turner, 2022
ISBN 9783110626285, e-ISBN 9783110626247

Organometallic Chemistry
Fundamentals and Applications
Ionel Haiduc, 2022
ISBN 9783110695267, e-ISBN 9783110695274

Process Technology
An Introduction
André B. de Haan and Johan T. Padding, 2022
ISBN 9783110712438, e-ISBN 9783110712445

Vahid Pirouzfar, Mastaneh Narimani,
Ahmad Fayyazbakhsh, Chia-Hung Su,
Milad Gharebaghi

Gasoline Additives

—

DE GRUYTER

Authors
Prof. Dr. Vahid Pirouzfar
Department of Chemical Engineering
Islamic Azad University
Central Tehran Branch
Ashrafi Esfahan Highway
Tehran, Iran
pirozfar@gmail.com

Ahmad Fayyazbakhsh
Faculty of Technology
Tomas Bata University
Nam. T.G. Masaryka 5555
760 01 Zlin, Czech Republic
Fayyaz_bakhsh@utb.cz

Milad Gharebaghi, M.D.
Tehran University of Medical Sciences
Teheran, Iran
miladgharabaghi@gmail.com

Mastaneh Narimani
Department of Chemical Engineering
Islamic Azad University
Central Tehran Branch
PO Box 19585/936
Tehran, Iran
mastanenarimani10@gmail.com

Chia-Hung Su
Ming Chi University of Technology
Gongzhuan Road
Taishan
Neu-Taipeh, Taiwan
chsu@mail.mcut.edu.tw

ISBN 978-3-11-099995-2
e-ISBN (PDF) 978-3-11-099996-9
e-ISBN (EPUB) 978-3-11-098983-0

Library of Congress Control Number: 2022942704

Bibliographic information published by the Deutsche Nationalbibliothek
The Deutsche Nationalbibliothek lists this publication in the Deutsche Nationalbibliografie;
detailed bibliographic data are available on the Internet at http://dnb.dnb.de.

© 2022 Walter de Gruyter GmbH, Berlin/Boston
Cover image: Petmal/iStock/Getty Images Plus
Typesetting: Integra Software Services Pvt. Ltd.
Printing and binding: CPI books GmbH, Leck.

www.degruyter.com

Preface

The present book represents a wide range of additives for gasoline used so far. As gasoline is among the most popular fuel (due to its price and energy), it is important to make its performance as much as possible. The main drawback of this fuel is that it contributes to most of the air pollution as well as greenhouse gases in the world. In this case, studying different used techniques to reduce air pollutants is important as eliminating gasoline use is still challenging. It should be noted that using gasoline additives should not negatively influence gasoline properties and its power. Thereby, this book studied the effects of different additives on the gasoline properties and power to make research comprehensive.

Consequently, this book aims to show the progress of additive usage in the gasoline industry to reduce air pollution and greenhouse gas emissions, enhance gasoline-powered engine performance, and improve gasoline properties such as octane number. Finally, a brief suggestion regarding future research has been made in the last chapter.

https://doi.org/10.1515/9783110999969-202

Contents

About the authors

Vahid Pirouzfar earned his Ph.D. in Chemical Process Engineering, process design, simulation, and control from Tarbiat Modares University Tehran, Iran. He has been working as a faculty member at the Islamic Azad University (IAU) since 2011, and currently he is a faculty member and Associate Professor at IAU, Central Tehran Branch.

Since 2007 he has been working in large companies, refineries, consultant and contractor research institutes such as Oil and Energy Industries Development Company (OEID), Central Petrochemicals (four big petrochemical complexes producing ethylene, HDPE, LDPE, LLDPE), Energy Renovation and Reclamation Consultants (Mabna), Iran Industrial Consultants Company, Middle East Energy Development Engineers Company (M.E.D. Co.), also the Oil and Energy Research Institutes in the fields of Oil, Gas, Petrochemical and Energy in Upstream and Downstream Industries, with various occupations as expert, master, supervisor, head of project and engineering department and managing director. He cooperated with international reputable and important companies like Shell and Sinopec during his activity.

V. Pirouzfar (*Ph.D.; Control, Simulation and Design of Chemical Processes Department*)
Department of Chemical Engineering,
Islamic Azad University, Central Tehran Branch,

P.O. Box:	14676–8683, Tehran, Iran,
Tel:	+98-912-2436110,
e-mail:	v.pirouzfar@iauctb.ac.ir
	pirozfar@gmail.com
Google scholar:	https://scholar.google.com/citations?user=ktSOIHgAAAAJ&hl=en
ResearchGate:	https://www.researchgate.net/profile/Vahid_Pirouzfar
ORCID:	http://orcid.org/0000-0002-2862-008X ✓
Linkedin:	https://www.linkedin.com/in/vahid-pirouzfar-0273b137
HomePage:	http://v-pirouzfar-chemeng.iauctb.ac.ir/faculty/en

Google scholar page: **ORCID:**   **Personal page:**

Mastaneh Narimani Bachelor of Chemical Engineering: Islamic Azad University Central Tehran Branch. Researched environmental impact of fossil fuels. To pursue her Master's degree in the field of Applied & Environmental Geoscinece in Tübingen university.

Bachelor of Chemical Engineering, Islamic Azad University, Central Tehran Branch, Ashrafi Esfahan Highway, Tehran Iran
Mastanenarimani1012@gmail.com,
(+98)912-022-3323

Ahmad Fayyazbakhsh had earned his Master's degree in Chemical Engineering before attending Tomas Bata University in 2020 as a Ph.D. candidate in Environmental Protection Engineering. Ahmad has published several articles and a book in this field. Moreover, he has three under-review articles (cooperative articles with different universities such as Yale and Arizona State Universities) on this subject.

Ph.D. candidate in Environmental Engineering, Tomas Bata University in Zlin, Faculty of Technology, Nam. T.G. Masaryka 5555. Zlin. Czech Republic
Fayyaz_bakhsh@utb.cz, (+420)776-847-055

Chia-Hung Su received his Ph.D. in Chemical Engineering from National Tsing Hua University in 2007. In 2009, he joined Ming Chi University of Technology where he is now Professor and Head of the Department of Chemical Engineering. Dr. Su has published over 100 technical papers, and his teaching and research interests are in the areas of process systems and control engineering for microbial, biochemical, and complex chemical processes.

Department of Chemical Engineering,
Ming Chi University of Technology,
Gongzhuan Road,
Taishan,
Neu-Taipeh. Taiwan
chsu@mail.mcut.edu.tw, (+886) 933 732 616

Milad Gharebaghi graduated in 2019 from Tehran University of Medical Sciences. He has made various contributions to multiple academic projects including composure, analysis, revision, and translation. Milad has worked in Hirmand which is affected by an extreme climate. There, he has examined and helped patients with environment-related illnesses and hydrocarbon poisoning. His research interests include occupational and environmental medicine.

M.D., Tehran University of Medical Sciences, Iran
miladgharabaghi@gmail.com,
(+98)920-3524353

1 History of gasoline

1.1 Introduction

Gasoline is made up of various chemical compounds and several additives, serving as an appropriate vehicle fuel. It is made from crude oil, which is a mixture of several distinct hydrocarbons in certain regions of the globe via refining. Crude oil consists of different products that differ from place to place, but gasoline forms roughly half an oil barrel in the United States [1, 2].

Furthermore, gasoline is available in a variety of forms or categories. Undiluted gasoline is the portion of the petroleum pond generated solely by crude oil distillation. Distillation is considered the primary refining stage. This step, however, does not produce appropriate fuel for the engines since gasoline's composition differs prominently in this stage. Crack gasoline, derived from the thermal or catalytic breaking of heavier oil fractions, accounts for the majority of gasoline used in automobiles and aviation. In this step called catalytic reforming, the formation of aromatic compounds increases the naphtha fraction's octane number. Also, further isomerization, which results from catalytic reforming, alkylation, isomerization, and polymerization, is conducted in the refining process to increase antiknock characterization. Finally, caustic washing, mercaptans oxidation, and desulfurization are carried out to make a suitable fuel. Undiluted gasoline, gasoline with cracks, reformed and synthetic gasoline, and supplements are used to create a wide range of gasoline [1, 3–5].

Gasoline mainly contains aromatic compounds, including alkylbenzenes as it is most abundant compound. Moreover, essentially all classes of hydrocarbons ranging from C_4 to C_{12} form gasoline [3]. The final gasoline's average molecular weight varies from 60 to 150 g/mol [5]. For better performance, additives could be added to gasoline as automobile fuel. Common additives are oxygenates and octane boosters, which are usually in much higher concentrations in the gasoline blend than other additives [1].

Gasoline is the primary transportation fuel in the United States: More than 25% of all energy usage in the US has been attributed to vehicle fuels [6]. Gasoline's energy capacity fluctuates between 112,500 Btu/gal and 114,000 Btu/gal depending on the season. The former defines the average capacity in the cold season and the latter in the hot season [7]. The energy content of standard gasoline fluctuates by as much as 3% to 5% between the lowest and highest energy levels from one batch and one station to another.

Although petroleum is used today to produce gasoline, coal has already been considered a powerful source. Until the beginning of the twentieth century, Germany depended almost entirely on imported crude oil for its fuel requirements. Thanks to scientific progress, high-pressure coal hydrogenation before World War I availed Germany and other European nations to generate substantial amounts of

https://doi.org/10.1515/9783110999969-001

gasoline [8]. However, coal could not compete with the readily accessible and easily refined petroleum-based motor fuels because it is a considerably more arduous and costly procedure to turn solid coal into liquid engine fuels.

Nowadays, to discover and carry petroleum, companies dig crude oil, refine it to produce gasoline, and sell it to a network of dealers who are experts in particular brand names and fuel blends. These large refiners may also offer these products to wholesalers, service stations, and other refiners.

In the combustion chamber, the optimal chemical reaction of fuel creates carbon dioxide and water. Nonetheless, unburned hydrocarbons, carbon monoxide, soot, nitrogen oxides, polycyclic aromatic hydrocarbons, and if the fuel includes sulfur or heavy metals, their oxides are produced as a consequence of the transient combustion processes. Since particle emissions significantly influence the environment and human health, their guidelines have principally focused on the particulate mass and number of gasoline direct injection engines worldwide. It has been shown that additives and physicochemical characteristics can affect octane number. For example, adding ethanol reduces emissions and lowers the potential risk of further health-related issues. In order to decrease hazardous emissions, today's gasoline has developed into a high octane, low sulfur fuel with the addition of oxygenates and restriction of maximum vapor pressure [9–12].

1.2 History of gasoline

Petroleum and its derivatives were known to several inhabitants in Eurasia, Africa, and the Americas, where it leaked to the surface and was utilized since at least 70,000 years ago: Hunters used bitumen to adhere spearheads to wooden shafts, settlers managed to apply petroleum in building constructions and vehicles, waterproofing canals, killing lice, and preserving mummies [13].

Although primarily recognized in the last two centuries, petroleum origin ages for millions of years. The organic material of most hydrocarbons has been derived from planktonic entities such as diatoms, blue-green algae, and foraminifera which were abundant about 541 million years ago. When these entities were buried in fine-grained sediments, which then would be called protopetroleum, they could be practically conserved. Going through a handful of physical and biochemical alterations might turn them into compounds, including kerosene, methane, and petroleum. As it can be presumed, oil has not been distributed evenly around the world. The Americas, Africa, Russia, and the countries which once formed the Soviet Union follow the proportion of reserves in the Middle East, which contain approximately half the globe's resources [14].

Petroleum usage as a tradable natural resource achieved momentum in the first half of the nineteenth century as Michael Faraday, in 1825, defined "new compounds of carbon and hydrogen" via heating oil, and Benjamin Silliman Jr. investigated the possibility of using petroleum as illumination substance. In the late nineteenth century, fuel acquired via coal and mineral oil could not resolve the global demand. Petroleum was on its dawn to dominate the global fuel market and transform civilization: new values and opportunities were to be established, advanced channels of science and technology were to be introduced, and modern approaches for energy generation and consumption were to be developed. The history of human civilization was turning over a new leaf, trying to improve the quality of life significantly [15, 16].

Before 1863, when commercial development of gasoline began, it was regarded as a futile side product of refining crude oil [12]. Before the electric lamp, kerosene was the primary oil refinery product for lighting, cooking, and heating. It has been suggested that gasoline had been isolated as a result of further kerosene refining in the 1860s. Primarily, gasoline was used in air-gas machines that could be burned to illuminate establishments. During the end of the nineteenth century, almost all gasoline production was utilized as a dissolvent by industries and dry-cleaning companies. When Nicolaus August Otto invented the first four-stroke cycle engine in 1876 Germany, gasoline was used as fuel [17].

The views about gasoline were changed since the beginning of the twentieth century when vehicles started dominating gasoline consumption. In 1919, internal combustion engines were using 85% of the nearly 90 million casks of gasoline in the United States in various vehicles. From 1899 until 1919, gasoline prices soared by more than 135%, from 10.8 cents per gallon to 25.4 cents per gallon, as demand surged. In 1929, 735,000 barrels were being consumed for aviation fuel. However, the fuel usage increased to over 6.4 million casks at the outbreak of World War II. Between 1929 and 1941, passenger automobile gasoline consumption climbed by more than 30 million barrels, gasoline was almost exclusively utilized to power automobile and aviation engines, and the amount of gasoline reached more than 50% of oil-based goods in general [4].

Gasoline was to be discontinued to be regarded as merely a by-product. More kerosene was refined to gasoline so that the demand could be met. New methods were being invented to ameliorate refinery efficiency: Thermal cracking, which used heat and pressure, was introduced in the 1910s and made it possible for unsuitable heavy petroleum products – that were more volatile – to be modified into compounds boiling in the gasoline range. Another process, thermal reforming, was developed in the 1930s to improve octane quality. Catalytic cracking was created in the late 1930s to shrink hydrocarbon molecules, in which a catalyst enabled the procedure at lower temperatures. Despite the invention of catalytic cracking, coking, an extreme form of thermal reforming, has continued to be widely used worldwide to refine heavy crudes or oils. In 1940, catalytic reforming, a method that uses ring closure to convert low-octane paraffin into higher-octane aromatics, came next. Other techniques which are used in refineries to boost octane include polymerization, alkylation, isomerization, and hydrocracking [12].

Since the 1950s, oil has become the world's most crucial energy source mainly due to its high energy content. In the United States, per capita, gasoline consumption climbed from around 150 gallons per year to just under 500 gallons per year between 1948 and 1975. Figure 1.1 shows gasoline consumption in the United States from the 1950s [18, 19]. The Middle East and Latin America established themselves as major oil producers by the 1970s. The cartel founded by the main Middle Eastern oil producers, known as OPEC (Organization of the Petroleum Exporting Countries), became a dominating player in global oil price-fixing by managing oil production. Even though several European countries such as Germany and France have not availed themselves of conventional oil production, they were able to develop methods to harness alternative energy sources, thanks to technology and creativity. Following the discovery of oil and gas under the North Sea in the 1960s, Britain, some Scandinavian countries, and the Netherlands became significantly self-sufficient in oil. In the 1980s, French and German industries started favoring diesel automobiles. It has been suggested that the reason behind it traces back to the 1973 and 1979 oil crises because when the crises subsided, Europe's diesel supply was far greater than the demand. Having said that, diesel engines emitted far less CO_2 than gasoline engines in the 1990s. However, this divergence between CO_2 production in the two types of engines has been considerably lowered as diesel engines have progressed and become more powerful than gasoline engines [20]. As offshore oil production started to fall in the 2000s, the European

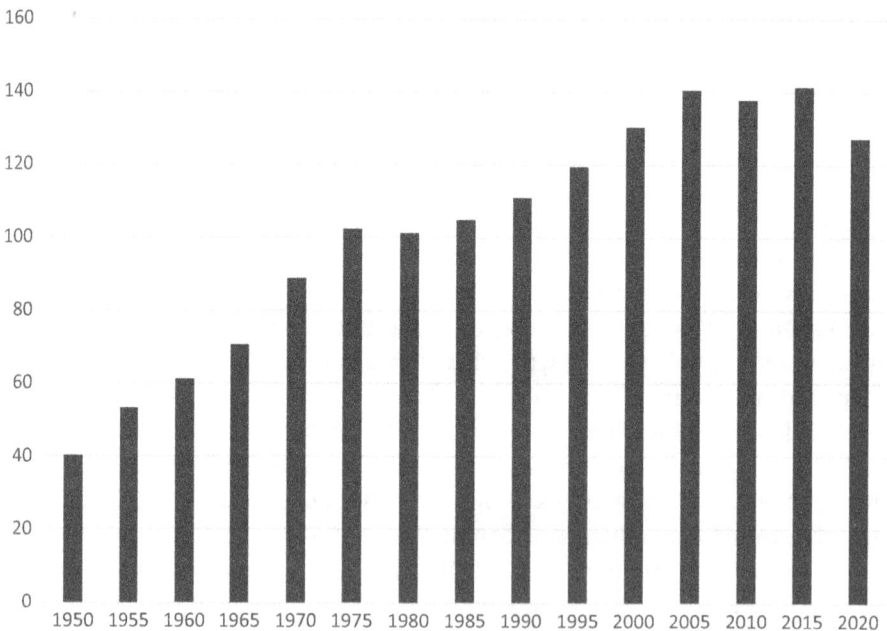

Figure 1.1: US motor gasoline consumption from 1950 to 2020.

Union has been inclined to more innovative and sustainable approaches, especially in the last decade, trying to be a petroleum-free society. This means lesser use of gasoline and diesel while more consumption of alternative energy forms [4, 16, 21].

Even so, the gasoline engines are here to stay inside the cities due to their ease, cost, and diversity, enabling them to work at higher speeds. Although the demand has decreased, or at most become stable in Europe and North America, gasoline demand has risen prominently in the rest of the world. The gasoline demand mainly rises from the growing middle class in the developing world. It has been expected that global oil products will grow by an average rate of 1.2% during 2012–2025. However, gasoline consumption growth leads to environmental changes as its combustion releases greenhouse gases, and its induced air pollution could be followed by health and medical issues [22, 23].

1.3 The influence of gasoline on World War I and World War II

After World War I, there was a wide range of variability in the kinds of fuels accessible on the commercial market. The variation was generally attributable to the origin of the crude oil as well as the techniques that were used to refine it. In order to transform the oil into gasoline, the development of the Octane Number Tests 219 was performed. Because fuels are composed of a combination of hydrocarbons, a significant amount of research was conducted in the early stages to determine the likelihood that specific hydrocarbon structures would knock. Ricardo undertook a series of experiments on various hydrocarbons in 1921, during which he discovered that certain arrangements of hydrocarbons are more susceptible to "knock" than others. The most effective antiknock capabilities were exhibited by aromatics, which are made of carbon atoms linked together in the form of a ring. In contrast, paraffin's antiknock performance, characterized by a continuous chain of carbon atoms, was the poorest. These discoveries were made with a degree of generalization; for instance, iso-octane, a paraffinic fuel, possesses outstanding antiknock qualities. It was discovered that Romanian gasoline included a high concentration of aromatics, which explains why it has such strong antiknock attributes [24]. Ethanol was the first additive of its kind to see widespread usage as an antiknock agent. Since the manufacturing method was well established at the time, ethanol saw widespread application as a motor fuel in early automobiles as well as airplanes. Furthermore, there was a political push for ethanol since it was believed it would benefit the farming sector. In 1906, researchers at the Edison Electric Testing Laboratory discovered that due to ethanol's antiknock qualities, it was able to function on compression ratios that were far greater than those required by gasoline. In 1918, due to the caustic character of ethanol, the high expense of producing it, and its low energy density, the focus on ethanol began to turn toward its use as an additive. Charles Kettering, who was the chief of research at General Motors (GM) at the time, contributed a significant amount of effort to this study. In 1918, he said that alcohol might be combined with gasoline to generate a viable motor fuel [25].

People were able to benefit from the antiknock technology that had been developed during World War II after the war. By 1944, oil companies were already speculating on the long-term consequences of the sudden surge in demand for high-octane fuels after the war. Because high-octane gasoline would be freely available at all gas stations after the war, GM predicted that postwar vehicles would have high-performance levels [26]. Fuel with an octane rating of 85 was readily available by 1945 when the average octane number was around 80. Because of the increasing abundance of high-octane fuel, the price of gasoline fell. Thirty dollars would buy a gallon of 100-octane gasoline in 1929. While the price had dropped to $16 in 1933, by 1943, it had fallen to just $0.20 for 100-octane fuel.

Ten years after the end of World War II, refining technology for high-octane fuel was allowed to expand over the world fully. The octane rating would be approximately 90, which is where it has been up to this point. Performance did not much increase with higher octane numbers.

Fuels with high octane values became widely available in the United States following World War II, influencing many automobile industry improvements. Compression ratio improvements led to an increase in engine power, which in turn resulted in a shift in consumer expectations for ever-increasing power. In response to these demands, the automobile industry got into a "horsepower race," creating "muscle cars" with ever-increasing power outputs. Ultimately, these so-called muscle cars would become synonymous with American automotive culture [27].

1.4 Gasoline mechanical and chemical properties

Gasoline, a blend of volatile and combustible liquid, consists of C4 to C12 compounds. Half of the motor gasoline is obtained from crude oil for engine use, which comes in two forms: aviation and automobile. Natural gasoline distillation ranges from 40 °C to 200 °C. The hydrocarbons formed at the boiling range are classified into aromatics, paraffin, and olefins. Interestingly, the characteristics of gasoline could differ based on the application and the refinery methods. Additionally, based on its octane number, motor gasoline is categorized into regular (85 to 88), midgrade (88 to 90), and premium (more than 90) gasoline [28, 29]. Table 1.1 shows the hydrocarbon components of refinery gasoline (IARC 1989).

Straight-run naphtha, hydrocracked naphtha, isomerate, reformate, alkylate, fluidized catalytic cracking, and polymer and pyrolysis gasoline build the most crucial body of gasoline. These components vary in volatility, chemical composition, density, and antiknock performance [28]. Chemists and mechanical engineers have spent much time researching the knocking process. The self-ignition of some unburned terminal gas before crossing by a typical propagating flame front is referred to as knocking.

Table 1.1: Typical composition of gasoline.

Compound		Weight percentage	Volume percentage
Olefins	Alkenes	2–5%	0–30%
	Cycloalkenes	1–4%	
Paraffin	Alkanes	4–8%	30–90%
Isoparaffin	Isoalkanes	25–40%	
Naphthene	Cycloalkanes	3–7%	
Aromatics		20–50%	10–50%
Benzene		0.5–2.5%	

Knocking occurs when the fuel's vapors burn quickly and prematurely in an engine's cylinders while the pistons are being compressed. This phenomenon emits a "metal ping" which is produced because of the pressure wave traveling across the cylinder chamber. The knocking can be assessed: the colliding and its reflection across the chamber wall make it possible to detect the pressure oscillation when the local pressure is being measured directly. It is usually between 4 and 20 kHz [30].

Prior to World War I, there was little research on knocking, and it was not until around 1920, when the size and power of automobile engines increased, that commercial attempts to solve the problem began. Knocking reduces power and efficiency, as well as causes engine damage over time. Nonetheless, standard criteria or scoring system did not exist until the late 1920s, when the octane number system was created to evaluate engine fuels. A fuel's octane number is a measurement of how likely it is to knock and has a minimum and maximum value.

Simple laboratory processes calculate the research octane number (RON). The fuels are burnt in the lab under precise and predetermined conditions. Normal heptane is an example of reference fuel. This is extremely low-quality gasoline with a 0-octane rating. 2,2,4-Trimethyl pentane (iso-octane) is at the other extreme of the spectrum that receives a 100 rating. To measure octane number, these two reference fuels are combined until they make some gasoline that "knocks" as much as the testing fuel. The chemical structures of these two have been demonstrated in Figure 1.2. Methanol, ethanol, and toluene exceed iso-octane in terms of

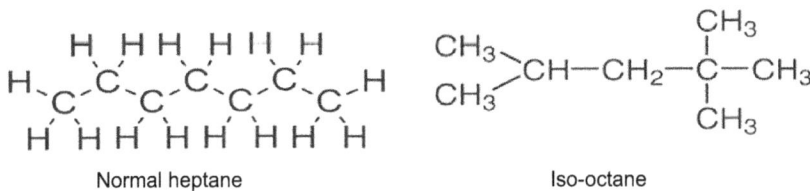

Normal heptane   Iso-octane

Figure 1.2: Chemical structures of normal heptane and 2,2,4-trimethyl pentane.

performance, and their octane ratings may be extrapolated to above 100. There is no doubt about it: the better the octane number, the more miles per gallon one gets out of their tank [31, 32].

Catalytic cracking technologies were created from the 1920s through the 1940s, which not only boosted gasoline processing efficiency but also gradually increased its octane number. Before the octane number system was invented, the Burton process, a non-continuous thermal process, was commercialized in 1913, producing gasoline with an estimated octane number of 50–60. Continuous thermal cracking, which dates more than 100 years, generates 73 RON gasoline. Cracked gasoline attained exceptional octane levels in the upper 80s thanks to the first catalytic process, Houdry technology (1938). Fluid catalytic cracking, the pinnacle of the cracking technique, was introduced in 1943 and increased the octane rating of gasoline to 95 [33].

When new octane-boosting chemicals and the octane number rating were introduced to the market in the 1930s, car fuel started being split into two groups: regular and premium, each with its octane range. Due to better refining processes and additives usage, gasoline's octane rating grew during the following 60 years. Additives' application became increasingly associated with environmental issues (such as clean air) and higher-octane ratings in the 1970s and 1980s. It is important to note that since the introduction of the original Ford Model T engine, gasoline has advanced significantly, and in constant dollars, it is currently less expensive [4, 34]. Some of the essential gasoline additives developed from the 1920s to the 1980s are summarized and shown in Table 1.2 and Figure 1.3. The addition of these compounds results in a rise in the octane number of gasoline.

Table 1.2: The octane number of common additives of gasoline.

Additive	RON
Ethanol	108
Methanol	107
Ethyl tert-butyl ether (ETBE)	118
Methyl tert-butyl ether (MTBE)	116
Tetraethyl lead (TEL)	100

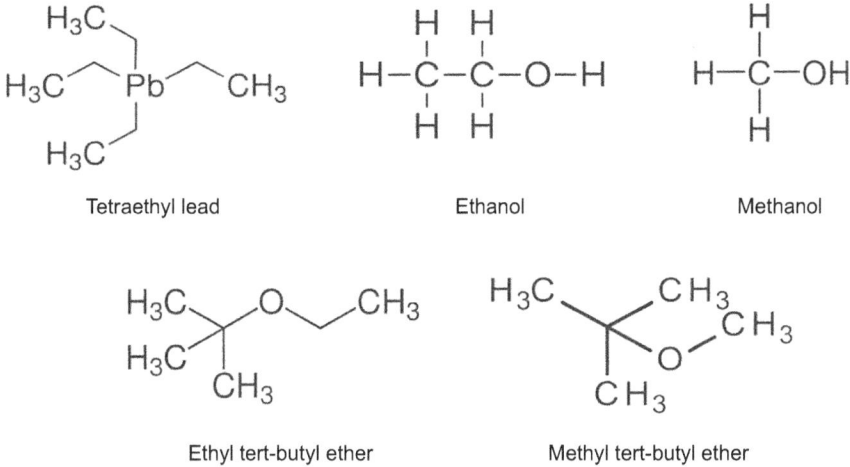

Figure 1.3: Chemical structures of some of well-known additives.

References

[1] Bergendahl J. Chapter 13 – Environmental Issues of Gasoline Additives – Aqueous Solubility and Spills. In: Letcher TM, ed. *Thermodynamics, Solubility and Environmental Issues*. Amsterdam: : Elsevier 2007. 245–58. doi:https://doi.org/10.1016/B978-044452707-3/50015-X

[2] Refining crude oil – U.S. Energy Information Administration (EIA). https://www.eia.gov/energyexplained/oil-and-petroleum-products/refining-crude-oil.php (accessed 25 Jun 2022).

[3] Stauffer E, Dolan JA, Newman R. CHAPTER 7 – Flammable and Combustible Liquids. In: Stauffer E, Dolan JA, Newman R, eds. *Fire Debris Analysis*. Burlington: Academic Press 2008. 199–233. doi:https://doi.org/10.1016/B978-012663971-1.50011-7

[4] Gasoline and Additives | Encyclopedia.com. https://www.encyclopedia.com/environment/encyclopedias-almanacs-transcripts-and-maps/gasoline-and-additives (accessed 26 Jun 2022).

[5] Owen K (Keith), Coley T, Weaver CS, *et al.* Automotive fuels reference book. Society of Automotive Engineers, Warrendale, Pennsylvania. 1995; 963.

[6] Use of energy for transportation – U.S. Energy Information Administration (EIA). https://www.eia.gov/energyexplained/use-of-energy/transportation.php (accessed 26 Jun 2022).

[7] Aceves S, Glaser R, Richardson J. Assessment of California reformulated gasoline impact on vehicle fuel economy. *Other Information: PBD: 1 Jan 1997* Published Online First: 1 January 1997. doi:10.2172/496233

[8] STRANGES AN. Synthetic Petroleum from High-Pressure Coal Hydrogenation. 1984; 21–42. doi:10.1021/BK-1984-0228.CH002

[9] Merker GP (Günter P), Schwarz Christian, Teichmann R. Combustion engines development: mixture formation, combustion, emissions and simulation. Springer. 2012; 642.

[10] Überall A, Otte R, Eilts P, *et al.* A literature research about particle emissions from engines with direct gasoline injection and the potential to reduce these emissions. *Fuel* 2015;**147**: 203–7. doi:10.1016/J.FUEL.2015.01.012

[11] Qian Y, Li Z, Yu L, *et al.* Review of the state-of-the-art of particulate matter emissions from modern gasoline fueled engines. *Applied Energy* 2019;**238**: 1269–98. doi:10.1016/J. APENERGY.2019.01.179

[12] Gibbs LM. How gasoline has changed. *SAE Technical Papers* Published Online First: 1993. doi:10.4271/932828

[13] Craig J. History of Oil: Regions and Uses of Petroleum in the Classical and Medieval Periods. 2020; 1–9. doi:10.1007/978-3-319-02330-4_38-1

[14] petroleum – Major oil-producing countries | Britannica. https://www.britannica.com/sci ence/petroleum/Major-oil-producing-countries (accessed 27 Jun 2022).

[15] Faraday M. On new compounds of carbon and hydrogen, and on certain other products obtained during the decomposition of oil by heat. *Philosophical Transactions of the Royal Society of London* 1825;**115**: 440–66. doi:10.1098/RSTL.1825.0022

[16] Craig J, Gerali F, Macaulay F, *et al.* The history of the European oil and gas industry (1600s–2000s). *Geological Society, London, Special Publications* 2018;**465**: 1–24. doi:10.1144/SP465.23

[17] Williamson HF (Harold F), Daum A. The American petroleum industry: the age of illumination 1859–1899. 1981; 864.

[18] U.S. domestic demand for gasoline 1990–2020 | Statista. https://www.statista.com/statis tics/188448/total-us-domestic-demand-for-gasoline-since-1990/ (accessed 28 Jun 2022).

[19] Chart: U.S. Gasoline Consumption Tripled Since 1950 | Statista. https://www.statista.com/ chart/1408/us-gasoline-consumption-tripled-since-1950/ (accessed 28 Jun 2022).

[20] Cames M, Helmers E. Critical evaluation of the European diesel car boom – Global comparison, environmental effects and various national strategies. *Environmental Sciences Europe* 2013;**25**: 1–22. doi:10.1186/2190-4715-25-15/TABLES/2

[21] Fridrihsone A, Romagnoli F, Cabulis U. Environmental life cycle assessment of rapeseed and rapeseed oil produced in Northern Europe: A Latvian case study. *Sustainability (Switzerland)* 2020;**12**. doi:10.3390/SU12145699

[22] Dehhaghi M, Kazemi Shariat Panahi H, Aghbashlo M, *et al.* The effects of nanoadditives on the performance and emission characteristics of spark-ignition gasoline engines: A critical review with a focus on health impacts. *Energy* 2021;**225**. doi:10.1016/J.ENERGY.2021.120259

[23] Landrigan PJ. Air pollution and health. *The Lancet Public Health* 2017;**2**:e4–5. doi:10.1016/ S2468-2667(16)30023-8

[24] Kovarik B. Henry Ford, Charles F. Kettering and the fuel of the future. *Automotive History Review* 1998;**32**: 7–27.

[25] Kettering CF. Studying the Knocks. *Sci Am* 1919;**125**: 364–364. doi:10.1038/ SCIENTIFICAMERICAN10111919-364

[26] Mittal V. The development of the octane number tests and their impact on automotive fuels and American society. *International Journal for the History of Engineering and Technology* 2016;**86**: 213–27. doi:10.1080/17581206.2016.1223940

[27] Abrams L, Knoblauch K. *Historians without borders: new studies in multidisciplinary history.* Routledge, an imprint of the Taylor & Francis Group 2020. https://www.routledge.com/Histor ians-Without-Borders-New-Studies-in-Multidisciplinary-History/Abrams-Knoblauch/p/book/ 9780367786557 (accessed 7 Jul 2022).

[28] Abdellatief TMM, Ershov MA, Kapustin VM, *et al.* Recent trends for introducing promising fuel components to enhance the anti-knock quality of gasoline: A systematic review. *Fuel* 2021;**291**. doi:10.1016/J.FUEL.2020.120112

[29] Ershov MA, Klimov NA, Burov NO, *et al.* Creation a novel promising technique for producing an unleaded aviation gasoline 100UL. *Fuel* 2021;**284**. doi:10.1016/J.FUEL.2020.118928

[30] Behrad R, Abdi Aghdam E, Ghaebi H. Experimental study of knocking phenomenon in different gasoline–natural gas combinations with gasoline as the predominant fuel in a SI

engine. *Journal of Thermal Analysis and Calorimetry* 2020;**139**: 2489–97. doi:10.1007/S10973-019-08579-W

[31] Anderson JE, Kramer U, Mueller SA, *et al.* Octane numbers of ethanol- and methanol-gasoline blends estimated from molar concentrations. *Energy and Fuels* 2010;**24**: 6576–85. doi:10.1021/EF101125C

[32] Waqas MU, Masurier JB, Sarathy M, *et al.* Blending Octane Number of Toluene with Gasoline-like and PRF Fuels in HCCI Combustion Mode. *SAE Technical Papers* 2018;**2018-April.** doi:10.4271/2018-01-1246

[33] Ayres RU, Ezekoye I. Competition and complementarity in diffusion: The case of octane. In: *Diffusion of technologies and social behavior.* Springer 1991. 433–50.

[34] Giebelhaus AW. Farming for fuel: the alcohol motor fuel movement of the 1930s. *Agricultural History* 1980;**54**: 173–84.

2 History of gasoline additives

2.1 Introduction

Fossil fuels serve as one of the crucial energy sources in the world. The petroleum sources are directly correlated with the increasing demand for energy production [1, 2]. The more fossil fuel sources deplete, and the population grows, the further the world goes on a path to an energy catastrophe. The most crucial fuel derived from petroleum is gasoline, which plays a key role in human life. Gasoline is mainly consumed in cars, sport utility vehicles, miniature aircraft, construction equipment and tools, farming, and so on. As gasoline demand increases gradually, its prices rise daily, and the fuel supply shortage puts many nations under economic and domestic pressure.

Moreover, consumption of fossil fuels leads to greenhouse gas buildup, acid rain, and ozone layer depletion. These hazardous effects impact environment negatively which puts the living in danger. Therefore, it is wise and necessary to hunt for alternative energy sources that can create appropriate energy, which are eco-friendly at a reasonable cost [3, 4]. Researchers have considered methods and additives to improve gasoline performance in recent decades [5, 6]. In the early twentieth century, research was being conducted to find a solution for fuels' "knocking" – a phenomenon occurring in engines – which produces high amounts of pressure that is problematic. So, many elements and compounds were being probed to be used as antiknock agents. In 1916, iodine as an additive to gasoline was suggested; however, Iodine was very expensive at the time. A year later, the alcohol and gasoline mixture was considered fitting fuel. The potential of an organometallic compound was found 5 years later in 1921 and that was unfortunately life-threatening as well: tetraethyl lead (TEL). Eventually, TEL was banned, and unleaded gasoline started to replace leaded gasoline [7].

Although scientists had previously claimed that TEL and methyl tert-butyl ether (MTBE) have a strong positive influence on the power of gasoline, their hazardous effects on the environment led researchers to work on alternative additives [8, 9]. In other words, organic lead (TEL) was employed as an antiknock additive in gasoline and jet fuels. The skin, lungs, and gastrointestinal system all absorb TEL quickly. It may be transformed into triethyl lead, which is hazardous. TEL accumulates in the limbic forebrain, frontal cortex, and hippocampus due to its highly lipophilic nature. Delirium, nightmares, irritability, and hallucinations are all symptoms of acute high-level exposure. Poor neurobehavioral scores in manual dexterity, executive function, and verbal memory are among the long-term impacts of TEL exposure. Most treatments focus on providing comfort; however, chelation therapy may be beneficial [10]. Also, MTBE is quickly becoming a serious environmental issue, but it might be studied by utilizing carbon and hydrogen isotope compositions.

The recent debate over the use of MTBE in gasoline to increase oxygen content and reduce carbon monoxide emissions to the atmosphere had resulted in a projected

https://doi.org/10.1515/9783110999969-002

phaseout of this chemical by 2002, albeit this has not yet occurred on a broad scale countrywide. MTBE has been incorporated at low quantities (2%) into various gasoline types to boost octane performance. Since 1979, it has been recognized as a carcinogen, although it was assumed to pose lower health risk than many other components in gasoline. MTBE has become a major pollutant of water and, to a lesser extent, air (when gasoline escapes from underground storage tanks and comes into contact with groundwater) (i.e., when distributed at a gasoline pump, however it is washed out relatively swiftly in the first rain). MTBE is also photodegraded quickly in the air. In many situations, MTBE dominates the chromatograms of MTBE-contaminated groundwater samples, with only modest quantities of other components present. Determining the actual source of MTBE can be problematic since many sites may have several sources of MTBE from many petrol stations in the region. Primary research suggested that the refractory character of MTBE gave a possible possibility for employing isotope-ratio mass spectrometry to determine the source/spill correlation of gasoline or MTBE. GCIRMS (Gas chromatography combustion isotope ratio mass spectrometry) is arguably the only practical approach that has any chance of discriminating distinct sources of MTBE in groundwater samples while not being the final answer [11].

Newer additives have been mainly classified into chemical-based and eco-friendly additives. Current commonly used chemical-based additives are demonstrated in Table 2.1. Eco-friendly or bio-additives, which are generally produced from plant oil or carbohydrate-rich feedstock (bio-alcohol), are especially practical in diesel engines that do not need modification in the engine itself. Moreover, they have a higher degradation rate and build less greenhouse gas. Palm oil has been indicated as a feasible additive for gasoline [3, 12–15].

Table 2.1: Major chemical-based additives.

Additive groups	Examples
Antiknock agents	
Oxygenates	
Alcohols	Methanol, ethanol, propanol, butanol
Ethers	Methyl tert-butyl ether (MTBE), ethyl tert-butyl ether (ETBE), tertiary-amyl methyl ether (TAME)
Organometallic compounds	Iron pentacarbonyl, ferrocene
Antioxidants	
Phenolic compounds	2,4-Dimethyl-6-tert-butyl phenol
Cetane improver	2-Ethylhexyl nitrate
Metal deactivators	N,N'-Disalicylidene-1,2-propanediamine

As presumed, additives are not limited to increasing octane agents. Diesel fuels usually generate foams when pumped into vehicles' tanks; antifoam agents such as polysilicon compounds overcome this issue; fuel system icing inhibitors come in handy in jet fuels while water tends to freeze in high altitudes; corrosion inhibitors are applied to prevent corrosion of steel equipment and infrastructure [16].

Some additives, which are usually used in other industries such as food preservation or cosmetics, have also been suggested as gasoline additives for fuel and engine stability. To illustrate, antioxidant additives inhibit some unfavored reactions resulting in fuel stability. A critical chemical reaction occurs regarding engine maintenance and oil storage: oxidation. Aromatics, olefins, and hydrocarbons, which all form gasoline, are highly susceptible to oxidation. Metal presence, light exposure, temperature rise, and oxygen make gasoline more prone to oxidation. This reaction builds a mass known as "gum" which affects gasoline properties and engine performance. The phenolic antioxidants have been studied since the 1960s and are able to thwart the auto-oxidation reaction. Metal deactivators inhibit oxidation, hydrocarbon-soluble salt production, and fuel thermal deterioration. Because of their antioxidant characteristics, they are sometimes considered a subgroup of antioxidant additives. They are employed in aviation gasoline and jet fuel. As for cetane improvers, the cetane number for diesel is somehow the same as the octane number for gasoline. The higher the diesel's cetane number, the fewer it would "knock" [17, 18].

The majority of greenhouse gases produced by industrial and transport activities contain more than 0.6 billion combustion units (especially cars), which is expected to be up to 2.5 billion units by 30 years [19–22]. The alternative fuel used as an additive must be derived from renewable sources. It is essential to use methods that do not affect the engine. Alcohol, especially ethanol and methanol, are the best choices in this regard [23, 24]. Studies on blending nanometals and alcohols with gasoline have proved that most of these additives are added to gasoline to reduce gasoline emissions and increase the fuel specifications such as octane number, brake power, and brake thermal efficiency (BTE). Within the last few years, several experiments have indicated that alcohol's influence is more robust than MTBE and TEL, especially from the environmental point of view [25]. Alcohols are subject to both reducing exhaust emissions and improving engine performance. The main factors of engine performance are BTE and brake-specific fuel consumption, which are enhanced by adding alcohol. However, some researchers have concluded that this additive may have a negative effect on air pollution and BTE. They have also come to the idea that increasing the percentage of oxygen in the blended fuel could prepare a condition for creating NO_x, CO, and CO_2 [26, 27]. Another advantage of blending alcohol with gasoline is the ability to meet the demand for higher octane numbers [28, 29]. Higher octane number allows the engine to operate at higher ratios of compression, and it causes higher thermal efficiency [30–32]. Some attempts have been made to blend nanoparticle with gasoline for different purposes [33–35]. There are different nano-additives with various potentials. The nano-additives in the emulsion have a high surface/volume ratio, which leads to better atomization and rapid evaporation [36–38].

Furthermore, some nanometals make the combustion more complete, which leads to a reduction in hazardous emissions [39, 40]. Moreover, some researchers used nanoparticles for methanol to gasoline (MTG), ethanol to gasoline (ETG) processes, gas to liquid, or acetone to gasoline, which are economical [41–43]. MTG and ETG are two critical chemical processes that effectively get beneficial chemicals from coal and natural gas by syngas [44, 45]. Table 2.2 illustrates the research on the influence of different additives on engine performance, exhaust emission, and engine performance. The oxygenated fuels affect gasoline properties (flash point, octane number, etc.). The engine load, injection pressure, engine speed, and other engine-related factors do not affect fuel properties.

This review paper aims to explain the influence of various additives on exhaust emissions, fuel properties, and gasoline engine performance by the latest research studies. The other purpose of this review is to investigate the impacts of blending many types of material and additives with gasoline fuel and compare each additive's influence.

Table 2.2: Chemical and physical properties of alcohols and gasoline.

Properties Material	Kinematic viscosity (cm^2/ s.10^{-6})	LHV (kJ/kg)	Octane No.	Boiling point (°C)	Flashpoint (°C)	Oxygenate molecules	Density (kg/m3)	Molecular weight (g/mol)
Gasoline	0.5	317.7	91–100	39	−46	0	700–800	
Methanol	0.75	1,100	108.7	64	11	50	790	32.04
Ethanol	1.2	840	110	78	13–14	34.8	788	46.07
n-Butanol	3	585	96	118	35–37	21.6	810	74.121

2.2 The first types of additives for gasoline

The engine of the car was one cause of banging. Knocking is more common in an automotive engine with a greater compression ratio, poor combustion, and detonation. The end-gas auto-ignition causes it before the flame front from the spark plug can reach it [46, 47]. Knocking happens when the compressed fuel–air combination bursts prior to igniting the spark plug. In high-compression internal combustion engines, gasoline is a chemical compound that is employed, and if it ignites excessively and bursts, it can harm the engine via knocking [48].

Researchers and scientists in the automotive industry discovered that knock-free engines were significantly more efficient and cost-effective during the first half of the twentieth century. This discovery was made due to extensive research and many experiments focusing on the problems that can arise in automobile engines [7]. Octane has a significant role in how resistant gasoline is to engine knocking;

hence, as gasoline's octane level rises, it becomes more resistant to engine knocking since it is more likely to be compressed by air and explode later. As a result, employing gasoline additives to raise the octane number of gasoline is one of the reasons why engines knock less when using fuel with a higher-octane number [49].

While working at the Dayton Engineering Laboratories Company (DELCO), which Charles F. Kettering had established in 1909, Charles F. Kettering was the one who discovered the impact problem. Thomas Midgley Jr., who was 27 years old at the time, became a member of the DELCO research staff in 1916. Midgley attended Cornell University for his education and received a degree in engineering there in 1911. After that, he found work at the Dayton location of the National Register Fund Company. After he had helped his father for the time being, he was away from DELCO for a while. Still, he eventually returned, and he and Kettering immediately began looking into the cause of the engine knock.

Knocking was not a significant issue with the low-compression vehicle engines of the day, but it hindered the development of higher-compression automotive engines that were both more efficient and powerful. The difficulty was that knocking got worse as the compression ratio increased, making it impossible to utilize higher compression ratios. If that had been possible, it would have led to better fuel efficiency and higher power output. Kettering's dream was to create an automotive engine that could operate at a more excellent compression ratio; hence, finding a solution to the knock problem was very necessary. It was assumed at the time that the knock was a result of preignition; however, Midgley's early work proved that it was generated by a rapid increase in pressure following the ignition [50–52]. Since 1916, many solutions have been proposed to minimize the damage to the vehicle's power plant in response to the challenges posed by engine knocking. Concurrently, Thomas Midgely started adding iodine to gasoline in order to avoid and lessen engine knocking. When added to gasoline as an additive, Iodine increases the fuel's octane level, which in turn results in a reduction in the amount of knocking placed on the engine. This conclusion was reached after an exhaustive battery of tests [7]. Even though it was beneficial to increase the octane number by adding iodine to gasoline, the practice was fraught with two significant drawbacks. The first issue was that iodine was very corrosive, which led to a great deal of damage being done to the engine's body. The second issue was that iodine was costly to utilize. Despite its inconvenience in usage, iodine's effectiveness as an antiknock agent spurred Midgley in his quest to find other agents. Because kerosene-soluble red dyes do not act as an antiknock, the red hue was quickly disproven and dropped. The antiknock activity was due to iodine, but colorless ethyl iodide was also an effective knock inhibitor; therefore, it was determined that iodine was the species.

The search for an efficient and practical antiknock agent was halted by World War I. Instead of searching for an antiknock compound that was both effective and practical, Midgley concentrated all of his efforts on inventing synthetic aviation gasoline. With the use of this approach, the hydrogenation of benzene was achieved in a

70:30 ratio, with the product being a combination of cyclohexane and benzene. As a result of the war, this could not be put to its intended use; nonetheless, its effectiveness was proven.

The Dayton Metal Products Company was founded in 1916 by Kettering after he sold DELCO the year before. The General Motors Company (GM) purchased the fledgling company's research division in 1919, and the subsequent GM Research Division was formed. At the time, Kettering was GM's vice president for research and oversaw this facility.

After serving in the military during World War II, Midgley returned to his quest for an effective antiknock chemical. With the help of Thomas A. Boyd and others, he tried to use all of the compounds that became available to him, but he was unable to accomplish it. After the merger was completed, he was given 2 weeks to find a new knock inhibitor to continue financing this research. GM would keep funding this study in the event of his success.

In 1919, Boyd discovered that aniline could be used as an effective antiknock agent. The timing could not have been better. Besides engineering, petrochemical and chemical research and development, as well as alternative fuels and engine design, needed to be improved. Since then, Kettering had built partnerships with firms like E.I. du Pont de Nemours and Standard Oil of New Jersey. Both companies were interested in antiknock agent research and development for reasons that should be self-explanatory [53]. The solution did not lie in aniline. Although it had beneficial antiknock action, it had a highly foul odor (which departed the exhaust), it oxidized in air, and there were worries about its toxicity as well as its corrosive effects on metals. In addition, the exhaust was filled with the smell of it. Since this was the case, aniline was eliminated from the running, and the search continued [54]. In 1917, Charles Kettering and Thomas Midgley participated in a research project together in which they tested the use of ethyl alcohol as an additive to gasoline. As a result of their findings, it was determined that using alcohol as an additive to gasoline is a preferable option because it is also friendly to the environment. They do not cause any damage, and the octane rating of the gasoline is raised as a result of their addition [7].

In the year 1854, a scientist from Germany made the initial discovery of TEL. A point of technical interest, but "its notorious deadliness" prevented it from being applied in commercial applications. It is perilous, and even light, prolonged contact with it has been known to induce hallucinations, breathing trouble, psychosis, spasms, paralysis, asphyxiation, and death in the most severe cases. Even light, prolonged contact with it has been known to cause these effects. In 1921, 67 years after it had been invented, it was still not being utilized; therefore, it was not an obvious choice as an addition to gasoline [55]. The search for new additives continued until 1921 when it was announced that TEL, a new additive, had been found by Thomas Midgley Jr. The production of this substance and additive in limited amounts were done in 1922, and the following year saw the first daring driver make use of it. To produce this additive, it is required to mix molten sodium and molten lead to create a highly reactive alloy. This may be done by combining the two metals in a melting pot. After that, it reacts with ethyl chloride to produce TEL [56, 57].

In 1924, GM laid the groundwork for what would become the Ethyl Corporation, with the intention of marketing lead-based gasoline. However, there were a few disadvantages to this particular gasoline. The toxicity of lead and the issues it causes for the environment are the first and most significant of these concerns. For instance, lead is responsible for a drop in IQ in children, as well as harmful effects on children's physical development [58], a poor influence on the immune system of birds [59], and the occurrence of acute malignancies in adults and rodents [60].

On the other hand, it shortens the engine's lifespan since it promotes inefficiency by depositing itself in various areas of the engine [56]. Lead, for instance, can form a deposit on the chamber of the combustion device, which entirely blocks combustion [56].

In 1925, a physiologist named Professor Yandell Henderson raised and cautioned the scientific and engineering communities about using TEL in gasoline as an additive. He did so for all of these reasons, but most importantly, for reasons related to medicine and public health, for the sake of humankind's survival [61]. But governments and scientific societies did not heed Professor Henderson's warnings. They continued to use this additive because very few scientific and engineering facilities and advancements were available before the 1940s. This made it difficult for them to detect the harmful effects of TEL on a small scale. Toxic compounds continued to accumulate up until the 1970s [62]. In the meantime, beginning in the 1940s and continuing onwards, with the expansion of human knowledge and access to radioactive materials, spectrometers for nuclear bombs began to be developed. These spectrometers could detect and geochemically study lead-induced contamination around the world [63].

However, in the United States, the use of TEL was not banned until 1973. After 1973, this additive was restricted over time, and in 1995, severe restrictions were implemented to prevent the use of this material, except for certain types of aircraft. In 2008, it was discontinued as an antiknock agent for usage in the market [64]. Also, in Europe, no legislation prohibited lead use until 1978. After that year, all EU nations were required to prohibit the manufacture of lead as well as its import [65].

2.3 TEL gets challenged by alternative additives

As was previously noted, one of the most significant reasons for utilizing TEL was to boost the octane number of gasoline to improve the engine's performance and act as an antiknock additive. The octane number is broken down into two distinct categories. The first group was called the research octane number . This number is connected to isooctane and n-heptane and is tested against them. The second classification is the motor octane number [54, 66]. It is possible to boost the octane number by utilizing additives, as well as the nature of the octane number itself, which is such that it rises with an increase in the number of carbon chain branches in a compound [67].

TEL was utilized as an octane booster and an antiknock agent in the 1920s and 1930s because it was inexpensive and effective. This made it the additive of choice during this period. However, viable replacements for TEL must be discovered because of its toxicity and environmental damage [67]. Even after TEL was made available on markets worldwide, large gasoline corporations continued their hunt for additional additives that may further lessen "knocking."

The addition of 1,2-dibromoethane helped reduce the deleterious effects of TEL and its tendency to precipitate. However, in order to accomplish this goal, considerable quantities of this chemical, which is both rare and costly, were required. At the same time, the Dow Chemical Company had wells in which they were able to extract bromine by adding chlorine to brines that already contained alkaline bromide [66]. After that, the factories of Ethyl and Dow joined forces and founded the Ethyl-Dow Chemical Company to exploit and manufacture bromine. During World War II, as a direct result of this collaboration, a sizable plant was built with the purpose of extracting bromine from seawater to use in the production of other chemicals [66]. But bromine was also one of the highly toxic elements. Because it was the primary component of chemicals, it was produced in large quantities until it caused unjustifiable environmental effects. Despite the fact that it took a very long time to raise awareness about the impact, it was produced in large quantities until it caused these effects [68]. As was previously mentioned, one of the issues that arose from the utilization of TEL was the accumulation of lead in the engine, which shortened the engine's lifespan. This issue was resolved by including ethylene dibromide, in the form of an additive, in the fuel that contained tetra-ethyl lead for some time. Ethylene dibromide released the lead in the engine into the atmosphere to safeguard the engine, which jeopardized the general population's health. Furthermore, methyl bromide and bromofluoro-compounds were responsible for causing damage to the ozone layer [68].

2.3.1 Methylcyclopentadienyl manganese tricarbonyl

Since TEL, Ethyl Corporation had not released any significant new antiknock chemicals until they released a manganese compound. They successfully identified the antiknock property of methylcyclopentadienyl manganese tricarbonyl (MMT) in 1954 and then released this product to the market. Manganese compounds were another type of antiknock material and octane booster additive that had a spike in popularity in the 1990s [69]. Despite the many benefits of this product and its fewer harmful effects than TEL, its use was limited, and oil companies in Europe were forced to voluntarily stop using MMT because of the many public concerns related to the use of MMT for human health and the environment. This was the case even though MMT has fewer harmful effects than TEL [69–74].

It was shown that MMT might be dangerous, especially to children. Although manganese was thought to have detrimental effects on the neurological and respiratory systems, and even though it created wear and deposits in the engine, it should be noted that manganese oxides had a negligible effect on the catalysts [75]. A federal appeals court in Washington, D.C., rejected the Environmental Protection Agency (EPA) ruling in 1998. The court's ruling authorized Ethyl Corp. to test MMT while selling it. There was no time limit for completing the exams in the legislation, and today MMT is being used in some countries as an additive [76].

2.3.2 Ferrocene

Ferrocenes emerged as more cost-effective substitutes for MMT in the effort to increase octane values. The most common ferrocene compounds in antiknock compositions are alkyl ferrocenes and dimethyl ferrocenyl carbinol. Both types of ferrocene compounds include alkyl groups [74].

Due to the iron-containing deposits produced from ferrocene, the use of iron compounds as antiknock compounds in Russia during the 1960s and 1970s resulted in forming a conductive coating on the surfaces of the spark plugs. This was a significant drawback of the practice [67]. In addition, wear and tear on an engine is accelerated due to ferrocene oxides being so rough. Iron oxide was discovered as a physical barrier between the exhaust gases and the catalyst/oxygen sensor. As a result, the catalyst began to erode and clog, making it less efficient as a catalyst [73, 77].

2.3.3 Oxygenated additives

Before 1990, environmental rules were inclined toward lowering emissions. Catalytic converters have long acted as powerful tools for combating air pollution. For the first time after 1990, laws attempted to change the gasoline composition. Since 1995, reformulated gasoline has altered the composition proportion via minimizing oxygen content and alteration of different contaminants. As additions, oxygenated compounds are used in several reformulated gasoline. The oxygen aids in the total combustion of fuel, lowering monoxide emissions.

2.3.3.1 Methyl tertiary butyl ether

The methanol derivative, MTBE, is the most common additive to deliver the extra oxygen needed to reformulate gasoline to meet these standards. This additive is currently found in approximately 30% of gasoline sold in the United States [78, 79].

In the 1940s, MTBE was initially utilized as a gasoline additive to replace lead in the 1970s and 1980s and was a popular additive in Europe. The introduction of MTBE as an octane-enhancing additive to gasoline in the United States began in 1979. It was primarily found in higher-grade gasoline and was only seldom found in lower-grade fuels. The most frequent usage of MTBE was found in New England, New Jersey, and other east coast states between 1979 and the mid-1980s. The usage of MTBE to raise gasoline's octane rating began to rise in the late 1980s. Following this, it was used to produce gasoline that burns more efficiently in certain conditions. As a result of the Clean Air Act of 1990's requirements, MTBE consumption soared across the country, eventually reaching epidemic proportions across most of the country. Due to stringent air quality regulations, MTBE was added to 95% of the gasoline sold in California in June 1996. Since then, it has become almost universally used in the state (1998) [29]. With the introduction of the federal reformulated gasoline program in the United States, MTBE consumption in the United States climbed dramatically between 1990 and 1995.

MTBE was heavily criticized in the late 1990s for its effectiveness and safety. According to a 1999 assessment by the National Research Council, adding oxygen additions to gasoline, such as MTBE and ethanol, is significantly less helpful in managing pollution than installing emission control technology and improving car engines.

Groundwater, lakes, and reservoirs have all been implicated in cases of significant sickness due to the presence of MTBE. The appearance of malignant tumors in some animal studies prompted many authorities to warn about the substance's potential health risks. Some institutes suggested that MTBE should be taken off the market as an additive for gasoline [80]. This started ethanol to find its place as an oxygenate additive in blending. Even so, today, more than 90% of the global population lives in areas that do not comply with the WHO Air Quality Guidelines [79].

2.3.3.2 Alcohol

When Kettering found that a TEL addition was unacceptable for usage, other corporations, such as Sun Oil Co., explored alternatives. For the first time, butyl alcohol additions to a gasoline blend based on highly fragrant crude petroleum were offered for sale by this firm [81]. Moreover, "White flash," an Arco product containing benzene, was on the market [82].

Since the early 1980s, ethanol, a fuel derived from corn, has been an increasingly prominent addition in the United States. Its usage originated in reaction to rising oil prices, as a gasoline alternative, and as a lead replacement for octane-enhancing purposes while lead was being phased out of the gasoline supply. More recently, its usage has been encouraged for reasons relating to the environment (the Clean Air Act Amendments). The primary function of fuel ethanol has shifted from a replacement for gasoline, which also played a limited role as an octane enhancer, to that of an additive that cleans the air. Before November 1992, ethanol was utilized as a gasoline replacement due to its market's peculiarities and the large subsidies offered by the federal government and individual states in the United States. It was used as a direct replacement for gasoline in a mixture called gasohol with 10% ethanol and 90% gasoline [83]. The most common mixture of gasoline and ethanol is called E10, consisting of 10% ethanol and 90% gasoline. In addition, ethanol can be used in its purest form, known as E100, or in a mixture known as E85, which is composed of 85% ethanol and 15% gasoline. Molasses is the primary component in the production of ethanol that is utilized in the United States. There is now research being conducted to determine the viability of commercializing the manufacture of ethanol derived from sources other than vegetables. These sources include cellulosic materials such as scrap wood or paper [84].

N-Butanol (C_4H_8O), on the other hand, has the potential to function either as a straight fuel or as an addition. Butanol, often known as "biobutanol," is a sustainable fuel when it is produced through the fermentation processes of a variety of microorganisms. Compared to ethanol, normal butanol offers many benefits that are widely recognized. These advantages include a more extensive energy content, more superb miscibility with transportation fuels and a lesser tendency to absorb water. Butanol, in its normal state, may be mixed with gasoline in any concentration, and because it has a larger energy content than a fuel generated from petroleum, it has to be blended in at a lower volumetric concentration to achieve the same level of performance. Because of these properties of butanol, there has been a resurgence of interest in how it burns [85–91].

When Germany relied on coal supply during World War II, methanol was used as a transportation fuel. There was a rush to find alternate motor fuels during the global energy crisis of the 1970s, which resulted in methanol-based fuels being employed in a broader range of applications later on. During this period, the state of California saw the most significant use of methanol. Around 21,000 M-85 specialized cars were put into service by the state under an M-85 fuel station scheme that used an M-85 combination of 85% methanol and 15% gasoline. State support for the M-85 program was discontinued when oil prices stabilized. Methanol has been allowed to be added to gasoline since the early 1980s, thanks to EN 228, a rule published by Comite Européenne de Normalisation, a European institution that sets standards. Refineries in Europe continue to use methanol and gasoline in their products, although only on a limited basis. The most widespread use of methanol–gasoline blends has occurred in China, where rising domestic methanol supplies, the majority of which are produced through the gasification of coal, as well as significant quantities imported from the

Middle East, have been used as an automotive fuel mix over the last decade. This is because China has the world's most enormous automotive population [92].

2.3.4 The latest additives in gasoline

The petroleum refining sector is primarily responsible for creating and mixing additives. Additives are critical to the industry's economic health since they increase gasoline and diesel fuel sales. Additives usually cost little more than three to four cents a gallon. Higher octane levels do not always mean lower fuel economy. They can be used for various purposes and may clean the engine and increase octane ratings, resulting in improved fuel economy.

Newer fuel additives do many tasks besides raising octane numbers and cutting emissions. For example, some increase gasoline's chemical stability, prolonging the storage life; others protect reservoirs, pipes, and vessels from corrosion and prevent deposit formation and dirt absorption; some prevent liquids from freezing, while others thwart foam creation when pumping [16].

Regulations to decrease automobile emissions into the environment have been established, so gasoline additives began to play an increasingly important role as an antipollution agent in the 1970s. Despite advancements in cracking and reforming methods, as well as additive research and blends (not to mention engine improvements), the usage of vehicle gasoline has increased air pollution. Modern distillates, which are mixtures of straight run and cracked or chemically modified products, have a greater aromatic concentration. A reason for increased particle and oxidative reactions arises from incomplete combustion.

Both the oil sector and automakers will be required to take potentially costly sulfur-reduction activities as a result of the regulations proposed by the EPA in the 1990s. These requirements have raised the need for fuel additives. Oxygenate additives such as alcohol are being used enormously. Other oxygenates such as ethers, esters, and carbonates are either being used or are under study to be used at the industrial level. The most rapidly expanding markets for fuel additives are non-premium gasoline and diesel fuel.

Other additives are being developed in addition to MTBE and MMT. Alcohol derivatives make up some of these. Ether derivatives, particularly ethyl-t-butyl ether, are also employed. Carbonates such as dimethyl carbonate are being considered to be used since it is regarded as "green" additive [93].

In addition to these, newer additives for gasoline are now being developed. These additives are designed to prevent carbon buildup, burned and deformed valves, high cylinder head temperatures, stuck valves, piston rings, clogged injectors, rough idle, and detonation.

2.3.5 Alternative fuels

Fossil fuel emissions such as NOx and CO have also been problematic for the environment and humans. However, it is notable that conversion to alternative fuels for economic reasons rather than environmental ones is not uncommon, while the petroleum industry has developed reformulated gasoline that burns significantly cleaner. Petroleum refiners should be able to meet increasingly rigorous environmental regulations for gasoline with only slight price increases in the future, thanks to predicted technological developments. Gasoline is the primary transportation fuel, but petroleum resources will decrease in time as their utilization and demand increase. Therefore, new alternative fuels should be researched and considered to produce enough energy nowadays and in the future.

The efficient substitutes would be those that could generate energy very efficiently, produce less pollution, and possess high stability and reliability. Keeping these in mind, fuel cells seem to be the cleanest method for generating sustainable energy now. Unlike battery cells, fuel cells can be continually supplied; therefore, their output may be maintained forever [94].

Hydrogen-based energy production has become a reality thanks to fuel cell development. Cars with hydrogen fuel cells are being developed worldwide. The Main hydrogen-based fuel cells that are being used are molten carbonate fuel cells, solid oxide fuel cells, and phosphoric acid fuel cells. Alkaline fuel cells and proton exchange membrane fuel cells are newer technologies that are needed to be studied more. The cost of producing electricity using hydrogen fuel cells is far lower than the power prices paid by the electric utility companies across the globe [95, 96].

When it comes to storing and transporting hydrogen, it is believed collecting it from a liquid source is the most efficient method. Additionally, creating gasoline reformers for fuel cell cars is a vulnerable and sensitive concept because of the gasoline sulfur level. Even a few parts per million may be enough to poison a fuel cell stack. Lastly, more public awareness is required due to the safety concerns associated with using hydrogen.

Microbial fuel cells (MFCs) generate protons and electrons, and their transportation and reaction with oxygen on the cathode generate electricity. *Geobacter* and *Comamonas* as microbial agents have been studied and showed great potential to decrease NOx emissions. Although MFCs' power output usually is lower than fossil fuels, and their equipment is not cheap, more advanced technologies might help them to gain higher efficiency [97, 98].

References

[1]   Khorramshokouh S, Pirouzfar V, Kazerouni Y, *et al*. Improving the Properties and Engine Performance of Diesel-Methanol-Nanoparticle Blend Fuels via Optimization of the Emissions and Engine Performance. *Energy and Fuels* 2016;**30**: 8200–8. doi:10.1021/acs. energyfuels.6b01856

[2]   Pirouzfar V, Fayyazbakhsh A. Diesel Fuel Additive. 2015.

[3]   Jadhav M, Jadhav S, Chavan S. Application of additives with gasoline fuel: A review. *E3S Web of Conferences* 2020;**170**. doi:10.1051/E3SCONF/202017001026

[4]   Garlapati VK, Mohapatra SB, Mohanty RC, *et al*. Transesterified Olax scandens oil as a bio-additive: Production and engine performance studies. *Tribology International* 2021;**153**. doi:10.1016/J.TRIBOINT.2020.106653

[5]   Kassem MGA, Ahmed AMM, Abdel-Rahman HH, *et al*. Use of Span 80 and Tween 80 for blending gasoline and alcohol in spark ignition engines. *Energy Reports* 2019;**5**: 221–30. doi:10.1016/j.egyr.2019.01.009

[6]   Kareddula VK, Puli RK. Influence of plastic oil with ethanol gasoline blending on multi cylinder spark ignition engine. *Alexandria Engineering Journal* 2018;**57**: 2585–9. doi:10.1016/ j.aej.2017.07.015

[7]   Nadim F, Zack P, Hoag GE, *et al*. United States experience with gasoline additives. *Energy Policy* 2001;**29**: 1–5. doi:10.1016/S0301-4215(00)00099-9

[8]   Ben H, Ragauskas AJ. One step thermal conversion of lignin to the gasoline range liquid products by using zeolites as additives. *RSC Advances* 2012;**2**: 12892–8. doi:10.1039/ c2ra22616b

[9]   Abdellatief TMM, Ershov MA, Kapustin VM, *et al*. Recent trends for introducing promising fuel components to enhance the anti-knock quality of gasoline: A systematic review. *Fuel* 2021;**291**: 120112. doi:10.1016/j.fuel.2020.120112

[10]   Schapira; AH V, Byrne E. *Neurology and clinical neuroscience*. 2007.

[11]   McGregor D. Fuel Oxygenates. In: Wexler P, ed. *Encyclopedia of Toxicology (Third Edition)*. Oxford:: Academic Press 2014. 671–81. doi:https://doi.org/10.1016/B978-0-12-386454-3.00025-7

[12]   Rusev DP, Kazakov PP, Iliev AL, *et al*. Use of bio additives in diesel engine fuels. *IOP Conference Series: Materials Science and Engineering* 2021;**1031**. doi:10.1088/1757-899X/ 1031/1/012014

[13]   Nair P, Meenakshi HN. Review on the synthesis, performance and trends of butanol: a cleaner fuel additive for gasoline. *International Journal of Ambient Energy* Published Online First: 2021. doi:10.1080/01430750.2021.1873849

[14]   Westphal GA, Krahl J, Brüning T, *et al*. Ether oxygenate additives in gasoline reduce toxicity of exhausts. *Toxicology* 2010;**268**: 198–203. doi:10.1016/J.TOX.2009.12.016

[15]   Yao C de, Zhang ZH, Xu YL, *et al*. Experimental investigation of effects of bio-additives on fuel economy of the gasoline engine. *Science in China*, Series E: *Technological Sciences* 2008;**51**: 1177–85. doi:10.1007/S11431-008-0170-1

[16]   Groysman A. Fuel Additives. *Corrosion in Systems for Storage and Transportation of Petroleum Products and Biofuels* 2014; 23–41. doi:10.1007/978-94-007-7884-9_2

[17]   Castro Dantas TN, Dantas MSG, Dantas Neto AA, *et al*. Novel antioxidants from cashew nut shell liquid applied to gasoline stabilization. *Fuel* 2003;**82**: 1465–9. doi:10.1016/S0016-2361 (03)00073-5

[18]   Namazifar Z, Saadati F, Miranbeigi AA. Synthesis and characterization of novel phenolic derivatives with the glycerol ketal group as an efficient antioxidant for gasoline stabilization. *New Journal of Chemistry* 2019;**43**: 10038–44. doi:10.1039/C8NJ05923C

[19] Dong G, Nagasawa H, Yu L, et al. Pervaporation removal of methanol from methanol/organic azeotropes using organosilica membranes: Experimental and modeling. *Journal of Membrane Science* 2020;**610**: 118284. doi:10.1016/j.memsci.2020.118284

[20] Pulyalina A, Rostovtseva V, Faykov I, et al. Application of polymer membranes for a purification of fuel oxygenated additive. Methanol/methyl tert-butyl ether (mtbe) separation via pervaporation: A comprehensive review. *Polymers (Basel)* 2020;**12**: 1–22. doi:10.3390/polym12102218

[21] Ershov MA, Potanin DA, Tarazanov S V., et al. Blending Characteristics of Isooctene, MTBE, and TAME as Gasoline Components. *Energy and Fuels* Published Online First: 2020. doi:10.1021/acs.energyfuels.9b03914

[22] Gorgich M, Mata TM, Martins AA, et al. Comparison of different lipid extraction procedures applied to three microalgal species. *Energy Reports* 2020;**6**: 477–82. doi:10.1016/j.egyr.2019.09.011

[23] Jin D, Choi K, Myung CL, et al. The impact of various ethanol-gasoline blends on particulates and unregulated gaseous emissions characteristics from a spark ignition direct injection (SIDI) passenger vehicle. *Fuel* 2017;**209**: 702–12. doi:10.1016/j.fuel.2017.08.063

[24] Tega DU, Nascimento H, Jara JL, et al. A Rapid and Versatile Method to Determine Methanol in Biofuels and Gasoline by Ambient Mass Spectrometry using a V-EASI Source. *Energy and Fuels* 2020;**34**: 4595–602. doi:10.1021/acs.energyfuels.9b02827

[25] Lim CS, Lim JH, Cha JS, et al. Comparative effects of oxygenates-gasoline blended fuels on the exhaust emissions in gasoline-powered vehicles. *Journal of Environmental Management* 2019;**239**: 103–13. doi:10.1016/j.jenvman.2019.03.039

[26] Rezvani MA, Shaterian M, Shokri Aghbolagh Z, et al. Oxidative desulfurization of gasoline catalyzed by IMID@PMA@CS nanocomposite as a high-performance amphiphilic nanocatalyst. *Environmental Progress and Sustainable Energy* 2018;**37**: 1891–900. doi:10.1002/ep.12863

[27] Costagliola MA, Prati MV, Florio S, et al. Performances and emissions of a 4-stroke motorcycle fuelled with ethanol/gasoline blends. *Fuel* 2016;**183**: 470–7. doi:10.1016/j.fuel.2016.06.105

[28] Sekhar C, Akimoto M, Fan X, et al. Jo ur l P re of. *Building and Environment* 2020;**184**: 107229. doi:10.1016/j.csite.2021.100891

[29] Anderson JE, Dicicco DM, Ginder JM, et al. High octane number ethanol-gasoline blends: Quantifying the potential benefits in the United States. *Fuel* 2012;**97**: 585–94. doi:10.1016/j.fuel.2012.03.017

[30] Ma Y, Yu Z, Wang Y, et al. Investigation on the influence of initial thermodynamic conditions and fuel compositions on gasoline octane number based on a data-driven approach. *Fuel* 2021;**291**: 120124. doi:10.1016/j.fuel.2020.120124

[31] Liu F, Zhou L, Hua J, et al. Effects of pre-chamber jet ignition on knock and combustion characteristics in a spark ignition engine fueled with kerosene. *Fuel* 2021;**293**: 120278. doi:10.1016/j.fuel.2021.120278

[32] Willems R, Willems F, Deen N, et al. Heat release rate shaping for optimal gross indicated efficiency in a heavy-duty RCCI engine fueled with E85 and diesel. *Fuel* 2021;**288**: 119656. doi:10.1016/j.fuel.2020.119656

[33] Bond TC, Doherty SJ, Fahey DW, et al. Bounding the role of black carbon in the climate system: A scientific assessment. *Journal of Geophysical Research Atmospheres* 2013;**118**: 5380–552. doi:10.1002/jgrd.50171

[34] Dastanpour R, Boone JM, Rogak SN. Automated primary particle sizing of nanoparticle aggregates by TEM image analysis. *Powder Technology* 2016;**295**: 218–24. doi:10.1016/j.powtec.2016.03.027

[35] Ali MKA, Xianjun H. Improving the tribological behavior of internal combustion engines via the addition of nanoparticles to engine oils. *Nanotechnology Reviews* 2015;**4**: 347–58. doi: doi:10.1515/ntrev-2015-0031

[36] Annamalai M, Dhinesh B, Nanthagopal K, *et al.* An assessment on performance, combustion and emission behavior of a diesel engine powered by ceria nanoparticle blended emulsified biofuel. *Energy Conversion and Management* 2016;**123**: 372–80. doi:10.1016/j. enconman.2016.06.062

[37] Ali MKA, Fuming P, Younus HA, *et al.* Fuel economy in gasoline engines using Al2O3/TiO2 nanomaterials as nanolubricant additives. *Applied Energy* 2018;**211**: 461–78. doi:10.1016/j. apenergy.2017.11.013

[38] Guerrero Peña GDJ, Hammid YA, Raj A, *et al.* On the characteristics and reactivity of soot particles from ethanol-gasoline and 2,5-dimethylfuran-gasoline blends. *Fuel* 2018;**222**: 42–55. doi:10.1016/j.fuel.2018.02.147

[39] Ranjan A, Dawn SS, Jayaprabakar J, *et al.* Experimental investigation on effect of MgO nanoparticles on cold flow properties, performance, emission and combustion characteristics of waste cooking oil biodiesel. *Fuel* 2018;**220**: 780–91. doi:10.1016/j.fuel.2018.02.057

[40] Soudagar MEM, Nik-Ghazali NN, Abul Kalam M, *et al.* The effect of nano-additives in diesel-biodiesel fuel blends: A comprehensive review on stability, engine performance and emission characteristics. *Energy Conversion and Management* 2018;**178**: 146–77. doi:10.1016/j. enconman.2018.10.019

[41] Shao J, Fu TJ, Chang JW, *et al.* Effect of ZSM-5 crystal size on its catalytic properties for conversion of methanol to gasoline. *Ranliao Huaxue Xuebao/Journal of Fuel Chemistry and Technology* 2017;**45**: 75–83. doi:10.1016/s1872-5813(17)30009-9

[42] Mei C, Wen P, Liu Z, *et al.* Selective production of propylene from methanol: Mesoporosity development in high silica HZSM-5. *Journal of Catalysis* 2008;**258**: 243–9. doi:10.1016/j. jcat.2008.06.019

[43] Han Z, Zhou F, Liu Y, *et al.* Synthesis of gallium-containing ZSM-5 zeolites by the seed-induced method and catalytic performance of GaZSM-5 and AlZSM-5 during the conversion of methanol to olefins. *J Taiwan Inst Chem Eng* 2019;**103**: 149–59. doi:10.1016/j. jtice.2019.07.005

[44] Chen H, Yang M, Shang W, *et al.* Organosilane Surfactant-Directed Synthesis of Hierarchical ZSM-5 Zeolites with Improved Catalytic Performance in Methanol-to-Propylene Reaction. *Industrial and Engineering Chemistry Research* 2018;**57**: 10956–66. doi:10.1021/acs. iecr.8b00849

[45] Xie Z, Liu Z, Wang Y, *et al.* Applied catalysis for sustainable development of chemical industry in China. *National Science Review* 2015;**2**: 167–82. doi:10.1093/nsr/nwv019

[46] Shi H, Tang Q, Uddeen K, *et al.* Effects of multiple spark ignition on engine knock under different compression ratio and fuel octane number conditions. *Fuel* 2022;**310**: 122471. doi:10.1016/J.FUEL.2021.122471

[47] Zhen X, Wang Y, Xu S, *et al.* The engine knock analysis – An overview. *Applied Energy* 2012;**92**: 628–36. doi:10.1016/J.APENERGY.2011.11.079

[48] Nguyen DD, Moghaddam H, Pirouzfar V, *et al.* Improving the gasoline properties by blending butanol-Al2O3 to optimize the engine performance and reduce air pollution. *Energy* 2021;**218**: 119442. doi:10.1016/J.ENERGY.2020.119442

[49] Zamankhan F, Pirouzfar V, Ommi F, *et al.* Investigating the effect of MgO and CeO2 metal nanoparticle on the gasoline fuel properties: empirical modeling and process optimization by surface methodology. *Environmental Science and Pollution Research* 2018;**25**: 22889–902. doi:10.1007/S11356-018-2066-3

[50] Beaton K. Professional Amateur: The Biography of Charles Franklin Kettering. By T. A. Boyd, with Introduction by Alfred P. Sloan Jr. New York, E. P. Dutton & Co., Inc., 1957. Pp. xii + 242. $4.50. *Business History Review* 1957;**31**: 445–7. doi:10.2307/3111423

[51] Seyferth D. The Rise and Fall of Tetraethyllead. 2. *Organometallics* 2003;**22**: 5154–78. doi:10.1021/OM030621B/ASSET/IMAGES/MEDIUM/OM030621BE00032.GIF

[52] Midgley T. *From the periodic table to production: the biography of Thomas Midgley, Jr., the inventor of ethyl gasoline and freon refrigerants*. Corona CA:: Stargazer Pub. Co. 2001.

[53] Garrett AB. Lead tetraethyl: Thomas Midgley, Jr., T. A. Boyd, and C. A. Hochwalt. *Journal of Chemical Education* 1962;**39**: 414–5. doi:10.1021/ED039P414

[54] Badia JH, Ramírez E, Bringué R, *et al.* New Octane Booster Molecules for Modern Gasoline Composition. *Energy and Fuels* 2021;**35**: 10949–97. doi:10.1021/ACS.ENERGYFUELS.1C00912/ASSET/IMAGES/MEDIUM/EF1C00912_0025.GIF

[55] Kitman JL. The secret history of lead. *NATION-NEW YORK-* 2000;**270**: 11.

[56] Nriagu JO. THE RISE AND FALL OF LEADED GASOLINE. 1990.

[57] Gibbs LM. Gasoline additives-when and why. *SAE transactions* 1990;**99**: 618–38.

[58] Zhou CC, He YQ, Gao ZY, *et al.* Sex differences in the effects of lead exposure on growth and development in young children. *Chemosphere* 2020;**250**. doi:10.1016/J.CHEMOSPHERE.2020.126294

[59] Vallverdú-Coll N, Mateo R, Mougeot F, *et al.* Immunotoxic effects of lead on birds. *Sci Total Environ* 2019;**689**: 505–15. doi:10.1016/J.SCITOTENV.2019.06.251

[60] Fu H, Boffetta P. Cancer and occupational exposure to inorganic lead compounds: A meta-analysis of published data. *Occupational and Environmental Medicine* 1995;**52**: 73–81. doi:10.1136/oem.52.2.73

[61] Mielke HW. Dynamic Geochemistry of Tetraethyl Lead Dust during the 20th Century: Getting the Lead In, Out, and Translational Beyond. *International Journal of Environmental Research and Public Health* 2018;**15**. doi:10.3390/IJERPH15050860

[62] Parsons PJ, McIntosh KG. Human exposure to lead and new evidence of adverse health effects: Implications for analytical measurements. *Powder Diffraction* 2010;**25**: 175–81. doi:10.1154/1.3402340

[63] Davidson CI. *Clean hands: Clair Patterson's crusade against environmental lead contamination*. Commack N.Y.:: Nova Science 1999.

[64] Abadin H, Taylor J, Buser MC, *et al.* Toxicological profile for lead: draft for public comment. 2019.

[65] Hagner C. Historical review of European gasoline lead content regulations and their impact on German industrial markets. 1999.

[66] Stratiev D, Kirilov K. OPPORTUNITIES FOR GASOLINE OCTANE INCREASE BY USE OF IRON CONTAINING OCTANE BOOSTER. *Petroleum and Coal* 2009;**51(4)**: 244–248.

[67] Demirbas A, Balubaid MA, Basahel AM, *et al.* Octane Rating of Gasoline and Octane Booster Additives. *http://dx.doi.org/101080/1091646620151050506* 2015;**33**: 1190–7. doi:10.1080/10916466.2015.1050506

[68] Rae ID. The Trouble with Bromine: Health and Environmental Impacts of Organobromine Compounds. *Global Environment* 2015;**7**: 106–33. doi:10.3197/197337314X13927191904880

[69] di Girolamo M, Brianti M di, Marchionna M. Octane Enhancers. *Handbook of Fuels* 2021; 403–30. doi:10.1002/9783527813490.CH14

[70] Weaver JB, Bates AG. WORKBOOK FEATURES COSTS – Inflation – How Far Is Up? *Industrial & Engineering Chemistry* 2008;**50**: 71A-73A. doi:10.1021/IE50586A008

[71] Walsh MP. The global experience with lead in gasoline and the lessons we should apply to the use of MMT. *American Journal of Industrial Medicine* 2007;**50**: 853–60. doi:10.1002/AJIM.20483

[72] Dabelstein W, Reglitzky A, Schütze A, *et al.* Automotive Fuels. *Ullmann's Encyclopedia of Industrial Chemistry* Published Online First: 15 April 2007. doi:10.1002/14356007.A16_719. PUB2

[73] Hilliard JC, Springer GS. Fuel economy in road vehicles powered by spark ignition engines; 453.

[74] Danilov AM. Fuel Additives: Evolution and Use in 1996–2000. *Chemistry and Technology of Fuels and Oils 2001 37:6* 2001;**37**: 444–55. doi:10.1023/A: 1014231230570

[75] Ross J. Fuels and Fuel-Additives. By S. P. Srivastava and J. Hancsók. *Energy Technology* 2014;**2**: 934–5. doi:10.1002/ENTE.201405008

[76] Davis JM, Jarabek AM, Mage DT, *et al.* The EPA health risk assessment of methylcyclopentadienyl manganese tricarbonyl (MMT). *Risk Analysis* 1998;**18**: 57–70. doi:10.1111/J.1539-6924.1998.TB00916.X

[77] Broch A, Hoekman K. Effect of Metallic Additives in Market Gasoline and Diesel. CRC Report No. E-114-2, https://crcao.org/reports/recentstudies2016/E-114-2 . . . 2016.

[78] Davis JM, Farland WH. The Paradoxes of MTBE. *Toxicological Sciences* 2001;**61**: 211–7. doi:10.1093/TOXSCI/61.2.211

[79] Ambient air pollution. https://www.who.int/teams/environment-climate-change-and-health/air-quality-and-health/ambient-air-pollution (accessed 3 Jul 2022).

[80] Yacobucci BD, Tiemann M, Mccarthy JE. CRS Report for Congress Renewable Fuels and MTBE: A Comparison of Provisions in the Energy Policy Act of 2005 (P.L. 109-58 and H.R. 6). 2006.

[81] Kovarik W. Ethyl-leaded gasoline: how a classic occupational disease became an international public health disaster. *Int J Occup Environ Health* 2005;**11**: 384–97.

[82] Little DM. *Catalytic reforming.* Tulsa Okla.:: PennWell Books 1985.

[83] Rask KN. Clean air and renewable fuels: the market for fuel ethanol in the US from 1984 to 1993. *Energy Economics* 1998;**20**: 325–45.

[84] Kim JS, Park SC, Kim JW, *et al.* Production of bioethanol from lignocellulose: Status and perspectives in Korea. *Bioresour Technol* 2010;**101**: 4801–5. doi:10.1016/J. BIORTECH.2009.12.059

[85] Harvey BG, Meylemans HA. The role of butanol in the development of sustainable fuel technologies. *Journal of Chemical Technology & Biotechnology* 2011;**86**: 2–9. doi:10.1002/JCTB.2540

[86] Jin C, Yao M, Liu H, *et al.* Progress in the production and application of n-butanol as a biofuel. *Renewable and sustainable energy reviews* 2011;**15**: 4080–106.

[87] Nigam PS, Singh A. Production of liquid biofuels from renewable resources. *Prog Energy Combust Sci* 2011;**37**: 52–68.

[88] Wallner T, Miers SA, McConnell S. A comparison of ethanol and butanol as oxygenates using a direct-injection, spark-ignition engine. *Journal of Engineering for Gas Turbines and Power* 2009;**131(3)**: 032802.

[89] Dürre P. Biobutanol: An attractive biofuel. *Biotechnology Journal* 2007;**2**: 1525–34. doi:10.1002/BIOT.200700168

[90] Shapovalov OI, Ashkinazi LA. Biobutanol: biofuel of second generation. *Russian Journal of Applied Chemistry* 2008;**81**: 2232–6.

[91] Szwaja S, Naber JD. Combustion of n-butanol in a spark-ignition IC engine. *Fuel* 2010;**89**: 1573–82.

[92] Netzer D, Antverg J, Goldwine G. Methanol proves low-cost, sustainable option for gasoline blending. *Oil and Gas Journal* 2015;**3**: 2015.

[93] Patil AR, Taji SG. Effect of Oxygenated Fuel Additive on Diesel Engine Performance and Emission: A Review. *IOSR Journal of Mechanical and Civil Engineering*; 30–5.www.iosrjournals.org (accessed 2 Jul 2022).

[94] Felseghi RA, Carcadea E, Raboaca MS, *et al.* Hydrogen fuel cell technology for the sustainable future of stationary applications. *Energies (Basel)* 2019;**12**. doi:10.3390/EN12234593

[95] Midilli A, Ay M, Dincer I, *et al.* On hydrogen and hydrogen energy strategies: I: current status and needs. *Renewable and Sustainable Energy Reviews* 2005;**9**: 255–71. doi:10.1016/J.RSER.2004.05.003

[96] Singla MK, Nijhawan P, Oberoi AS. Hydrogen fuel and fuel cell technology for cleaner future: a review. *Environmental Science and Pollution Research* 2021;**28**: 15607–26. doi:10.1007/S11356-020-12231-8

[97] Han X, Qu Y, Wu J, *et al.* Nitric oxide reduction by microbial fuel cell with carbon based gas diffusion cathode for power generation and gas purification. *Journal of Hazardous Materials* 2020;**399**. doi:10.1016/J.JHAZMAT.2020.122878

[98] Munoz-Cupa C, Hu Y, Xu C, *et al.* An overview of microbial fuel cell usage in wastewater treatment, resource recovery and energy production. *Science of the Total Environment* 2021;**754**. doi:10.1016/J.SCITOTENV.2020.142429

3 The usage of oxygenated additives in gasoline

3.1 Introduction

Oxygenates are organic compounds with oxygen atom(s) in their chemical structures. Being oxygen-rich enables them to burn cleaner and make less pollution. As removing lead as an antiknock agent from ordinary gasoline decreases its octane number significantly, other additions were required to raise the octane number when lead was prohibited. That brings us to applying oxygenates to increase fuels' antiknock trait. Oxygenates are added to both gasoline and diesel. When added to gasoline, they generally reduce air pollution while more complete combustion in engines is guaranteed. Even so, CO emission reductions are often less in modern cars and bigger in older vehicles. Applying oxygenates to diesel decreases harmful emissions via lowering NO_x and particulate projection. Since fuels blended with oxygenates reduce knocking and generate higher compression inside the engine, they increase the engine's horsepower. Moreover, oxygenates lead to a rise in fuel consumption, hence profitable to fuel economy. This result is due to a slight decrease in the energy content of mixed fuel. Table 3.1 shows some well-known oxygenates [1, 2].

Table 3.1: Oxygenates.

Oxygenates	Examples
Alcohols	Methanol, ethanol, propanol, butanol, propanol
Ethers	Methyl tert-butyl ether (MTBE), ethyl tert-butyl ether (ETBE), tertiary-amyl methyl ether (TAME)
Carbonates	Dimethyl carbonate (DMC), diethyl carbonate (DEC)
Esters	Methyl acetate, ethyl acetate

The two prominent families of oxygenated additives are alcohols and ethers, which will be discussed in this chapter and briefly describe other groups.

3.2 Ethers

One of the most common additives for gasoline are ethers such as methyl tert-butyl ether (MTBE), tert-amyl ethyl ether (TAEE), and di-tert-amyl ether (di-TAE). They have been used to improve the octane number and reduce soot and CO emissions. Ethers are preferred as gasoline additive over alcohols because their octane values are more outstanding, have lower vapor pressures, and are more predictable in

https://doi.org/10.1515/9783110999969-003

blending with gasoline. However, the usage of ethers shortened in the United States when the government mandated the use of ethanol [3].

3.2.1 Methyl tert-butyl ether

MTBE is one of the earliest oxygenates, first used as an octane enhancer to replace lead in gasoline and later used to minimize carbon monoxide emissions. It is mainly derived from methanol and isobutylene. Because of its low cost, high-octane value, and ease of inclusion into gasoline stock, MTBE was the primary option of gasoline oxygenate used globally beginning in the late 1970s [1, 4]. Despite claims that MTBE negatively influences the environment by poisoning water sources, it is widely used as an octane enhancer in motor gasoline throughout Europe, the Middle East, Africa, Asia, and Latin America [5]. Recent bio-based MTBE for gasoline break-throughs is projected to generate prospects for the MTBE industry [3].

3.2.2 Ethyl tert-butyl ether

Ethyl tert-butyl ether (ETBE) may be made by heating ethanol and isobutene together. It has been suggested as a substitute for MBTE. As an addition to fuel, ETBE has the potential to significantly reduce CO and unburned hydrocarbon (HC) emissions, al-though because of greater oxygen availability, NO_x emission increases slightly. ETBE is thought to be unharmful to health even at high concentrations [1, 6].

3.2.3 Tert-amyl ethyl ether

TAEE and di-TAE are high-molecular-weight branching ethers derived from largely renewable sources that are not commercially manufactured. However, they are pro-spective options for usage as high-purity fuels and in gasoline compositions. Thanks to the ability to increase octane number, they could be considered suitable additives for gasoline.

3.2.4 tert-Amyl methyl ether

tert-Amyl methyl ether (TAME) is a transparent flammable liquid soluble in ethers, alcohols, and HCs. It is widely used as an octane booster and to raise the oxygen content in gasoline. TAME has a research octane number of 112, but has a lower octane number compared to MTBE, making its progress in industrial settings slow [1, 7].

3.3 Alcohols

Alcohols play a crucial characteristic as fossil fuel additives, especially for diesel and gasoline. The benefits of alcohols and the purpose of their usage for gasoline and diesel could be different. To be more specific, ethanol has been used as an oxygenated additive for gasoline to improve its octane number; however, the purpose of this additive in diesel is entirely different. In this regard, the usage of alcohol in diesel is for environmental purposes, although in gasoline, it could be used environmentally or as an additive to improve its combustion factors.

Alcohols have hydroxyl groups with various carbon atoms. The properties of gasoline and alcohol are listed in Table 3.2. In this part, we provide some details about different alcohols and their properties. Ethanol has been preferred to other alcohols specifically for its properties, such as improved volatility, oxygen percentage, and latent heat properties [8, 9]. The researchers concluded that increasing the percentage of fuel oxygen can provide the conditions by which the CO, NO_X, and CO_2 can be enhanced [10, 11]. Some decades ago, alcohol like ethanol was found to be an essential fuel for engine combustion. Gasoline incorporated with up to 5–7% ethanol can be applied as a fuel for cars. It has been available for years in different markets such as the United States and Brazil [12, 13]. The influence of these additives on the environment and fuel properties will be addressed in the following parts.

3.3.1 Methanol

Methanol (CH_3OH) or methyl alcohol is a chemical substance produced when a methyl group is linked to a hydroxyl group. Figure 3.1 shows the worldwide production of methanol from 2018 to 2020 and a forecast for 2030 [27]. Methanol is also known as wood alcohol because it was once produced mainly by destructive wood distillation. Methanol is applied to be a significant blending agent for pure oil. Previous studies have reported that about 22% of methanol can be added to pure oil to run an engine without dangerous air pollution [11, 28]. Moreover, methanol is produced through a simple procedure. It is possible to produce methanol by a bed reactor using a supported catalyst with iron and copper nanoparticles [11].

As methanol is a mixture of hydrogen gas, CO_2, and CO, it is necessary to produce a gas obtained from any plant using general gasification technologies to synthesize [13]. Methanol will be applied as an energy source that develops suitable energy storage, HCs, and fuel. As an appropriate octane gasoline blend, methanol reduces carbon monoxide and HC emissions by improving engine compression [12, 26]. In methanol to HCs conversion on ZSM-5 zeolites, the ratio of HC compounds in the product can be considerably changed by altering reactants' pressures. Differences in the reactants' partial pressures can significantly shift the reaction's aromatization

Table 3.2: Influence of blending different additives on engine performance, exhaust emission, and octane number.

Research group	Oxygenated additives and other additives			Speed	Desulfurization rate	Exhaust emission				Fuel performance			Octane number
	Alcohol	% Alcohol	Other additives			UHC	NO_x	CO	CO_2	BSFC	BTE	BP	
Thakur et al. [14]	Ethanol	0	–	2,000	–	–	–	–	–	0.26	30	–	–
		5	–	2,000	–	–	–	–	–	0.263	31	–	–
		10	–	2,000	–	–	–	–	–	0.264	32.5	–	–
Amirabedi et al. [15]	Ethanol	10	–	2,800	–	−21.51%	–	−6.39%	–	−21.81%	–	2.63%	–
		10	–	2,800	–	−40.64%	−23.43%	−21.55%	–	−34.69%	–	14.38%	–
		10	–	2,800	–	−51.83%	−32.34%	−24.09%	–	−38.89%	–	19.56%	–
Badra et al. [16]	Ethanol	0	–	–	–	–	–	–	–	–	–	–	92
		10	–	–	–	–	–	–	–	–	–	–	95
		25	–	–	–	–	–	–	–	–	–	–	100
Ilhak et al. [17]	Ethanol	0	–	3,500	–	65	790 ppm	0.64%	16.30%	231.5	–	32.1	–
		10	–	3,500	–	66.5	720 ppm	0.56%	15.20%	238	–	32	–
		20	–	3,500	–	69	610 ppm	0.58%	13.50%	241.6	–	32.4	–

Al-Hasan [18]	Ethanol	0	-	2,000	0	0	0	0	450	-	-	-
		10	-	2,000	-7%	-21%	-8.50%	-5%	14%	-	-	-
		20	-	2,000	-17%	-29%	-13%	-7%	19%	-	-	-
Kassem et al. [19]	Ethanol	10	ST80	1,500	138 ppm	22 ppm	3.16 ppm	4.36 ppm	-	-	-	95
		20		1,500	123 ppm	19 ppm	2.48 ppm	4.52 ppm	-	-	-	95
		25		1,500	98 ppm	19 ppm	2.34 ppm	4.64 ppm	-	-	-	92
	Methanol	9		1,500	152 ppm	23 ppm	3.29 ppm	4.13 ppm	-	-	-	95
Jin et al. [20]	Ethanol	>0.01 wt%	-	-	-	0.002 g/km	0.275 g/km	190 g/km	-	-	-	92
		13.12 w.-%	-	-	-	0.005 g/km	0.292 g/km	187 g/km	-	-	-	94
		53.63 wt%	-	-	-	0.004 g/km	0.200 g/km	188 g/km	-	-	-	100
		87.8 wt%	-	-	-	0.003 g/km	0.147 g/km	187 g/km	-	-	-	106
Lim et al. [21]	Ethanol	3	-	-	0.46	0.36	0.62	-	-	-	-	-
		6	-	-	0.41	0.4	0.55	-	-	-	-	-
		10	-	-	0.39	0.48	0.56	-	-	-	-	-

(continued)

Table 3.2 (continued)

Research group	Oxygenated additives and other additives					Exhaust emission				Fuel performance			Octane number
	Alcohol	% Alcohol	Other additives	Speed	Desulfuri- zation rate	UHC	NO$_x$	CO	CO$_2$	BSFC	BTE	BP	
Iodice et al. [22]	–	–	–	–	–	1.48	0.156	8.4	298	–	–	–	–
	Ethanol	20	–	–	–	1.02	0.123	6.8	283	–	–	–	–
		30	–	–	–	1.28	0.117	7	262	–	–	–	–
Kareddula et al. [23]	Ethanol	5	15 PPO5E	1,500	–	28.48%	−0.41%	−24.11%	–	7.20%	−0.40%	–	–
	–	0	15 PPO	1,500	–	52.63%	−24.23%	−9.70%	–	−4.10%	9.35%	–	–
Geng et al. [24]	MMT8	8	–	2,000	–	−123.54%	8.08%	13.59%					94.4
	MMT12	12	–	2,000	–	−19.22%	7.53%	64.56%					94.5
	MMT18	18	–	2,000	–	−7.70%	11.20%	96.80%					94.9
Ashraf Elfasakhany [25]	Ethanol	7 vol%	–	3,450	–	200	–	2.9	12.6	–	–	2.35 + V44	–
		10	–	3,450	–	190	–	2	13.2	–	–	2.38	–
	Methanol	3	–	3,450	–	210	–	3.6	12.4	–	–	2.34	–
		7	–	3,450	–	215	–	1.5	13.6	–	–	2.24	–

Yilmaz et al. [26]	Methanol	5	–	2,000	–	–12%	1,449 ppm	–	–	–	10.30%
		5	Hydrogen 6	2,000	–	–	–	–	–	–	4.90%
		5	Hydrogen 15	2,000	–	–	–	–	12.06%	–	8.70%
		15	–	2,000	–	–40%	–	–	–	18.80%	–
		15	Hydrogen 6	2,000	–	–	–	–	–	–	6.70%
		15	Hydrogen 15	2,000	–	–	2,296 ppm	0.09%	–	–	11.70%

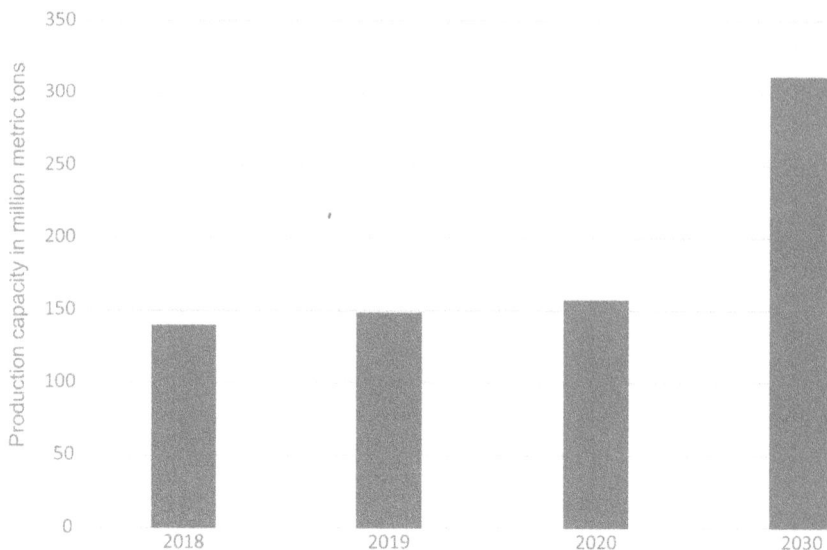

Figure 3.1: Production of methanol around the world with a forecast of 2030.

steps and dehydration relative rates [29]. The performance of synthesized samples cat-alytic in MTG (methanol-to-gasoline) reaction was studied by Wan et al. [30] in a fixed-bed reactor. They reported that although the catalyst activity was higher sustained with increasing SiO2/Al2O3 ratio, reducing SiO2/Al2O3 ratio caused higher methanol conversion. Firoozi et al. [31] experimentally studied the influence of ZSM-5 crystal size on the methanol to propene (MTP) process's performance. They concluded that during a 200 h testing reaction, the performance of nanosized ZSM-5 was better than micro-sized ZSM-5 in methanol conversion. The long lifetime and the high catalytic activity of the 36 small-sized ZSM-5 were attributed to more available acid sites.

3.3.2 Ethanol

Ethanol (also called ethyl alcohol) is a chemical compound that is often called EtOH. EtOH is typically produced by sugar fermentation, a well-known technique world-wide, especially in Brazil. Figure 3.2 shows the worldwide distribution of ethanol production in the last year [32]. Blending alcohols, especially ethanol, with gasoline has become commonplace partly because the octane number of ethanol is higher than gasoline [33–35]. Ethanol also has a higher vaporization latent heat than gaso-line [36]. Consequently, ethanol blending can supply excess consequent and cool-ing fuel knocks resistance and volumetric efficiency advantages in DI engines [37]. One of ethanol's benefits indicated in the previous research is ethanol to gasoline conversion (ETG). This process mainly dealt with catalyst development.

The catalytic dehydration and conversion of ethanol to propylene and ethylene are some results obtained from the previous research using ZSM-5 zeolite, which relies on catalyst acidity [38, 39]. The results of a study by Gayubo and his colleague [40] indicated that using ZSM-5 as a catalyst for ETG is highly beneficial due to the catalyst's stability against coking. Improving the diffusion property of the reactant and product molecules into and out of ZSM-5 is a practical strategy to decrease the deactivation rate caused by carbon deposition [41, 42]. Thus, ZSM-5 with mesopores triggers high catalytic stability. Viswanadham et al. [43] considered the performance of nanocrystalline H-ZSM-5 in the ETG process. They concluded that the nano-zeolite, including excess porosity and intense acidity, has shown enhanced gasoline production, rich in branched paraffin and aromatics. This gasoline possesses a low concentration of benzene (which reduces PAH), a high concentration of toluene, xylenes, isodecane, and enhances fuel applications' suitability.

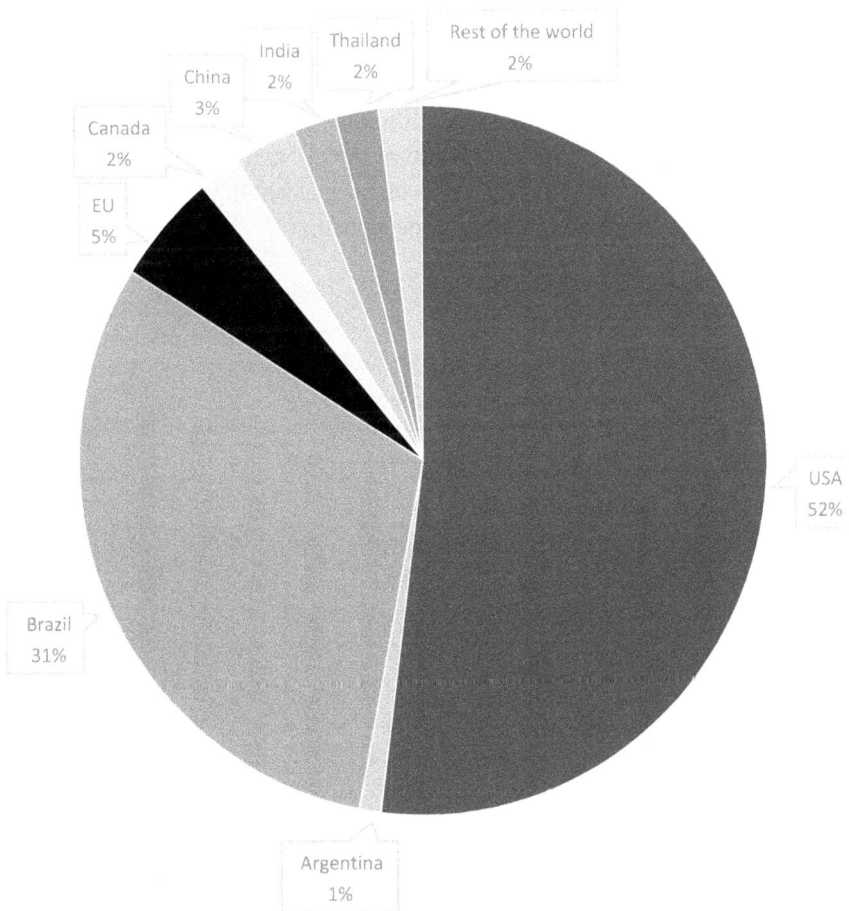

Figure 3.2: Distribution of ethanol production around the world in 2020.

3.3.3 Butanol

Although ethanol and methanol have several advantages, they cause corrosion. Thus, isobutanol is a better choice with a significant worldwide market, as shown in Figure 3.3 [44]. Since butanol absorbs less water and can be blended, it has more energy than ethanol and methanol. Butanol has become an alternative fuel for gasoline and diesel fuels. Butanol conversion into fuel is more costly and toxic, so its production rate is less than ethanol and methanol [45–48]. All butanol forms have equal energy, even though their physical properties are different. All forms of butanol can be used as fuel. Recently, most studies are dealt with the combination of butanol gasoline in internal combustion engines [49, 50]. In this regard, blending different types of butanol and gasoline in various engine rotations can increase and decrease pollutants such as CO, CO_2, and UHC according to the conditions of the engine like being regulated or not-regulated [51].

The fact is that any fuel with high viscosity deposits carbon in engine combustion [52, 53]. Butanol enhances the viscosity of ethanol and gasoline blending [54]. The enhancement of fuel viscosity by additives causes increase in fuel injection time [55]. Using a minor percentage of butanol slightly increases the viscosity of the fuel [56].

Regarding the properties of n-butanol, this substance can enhance the performance and the co-efficiency of the exploitation in cold-start conditions. Previous studies [57–59] indicated that the emission of normal butanol pollutants purely produces more HC than gasoline although decreasing CO_2 and NO_x. Also, engine

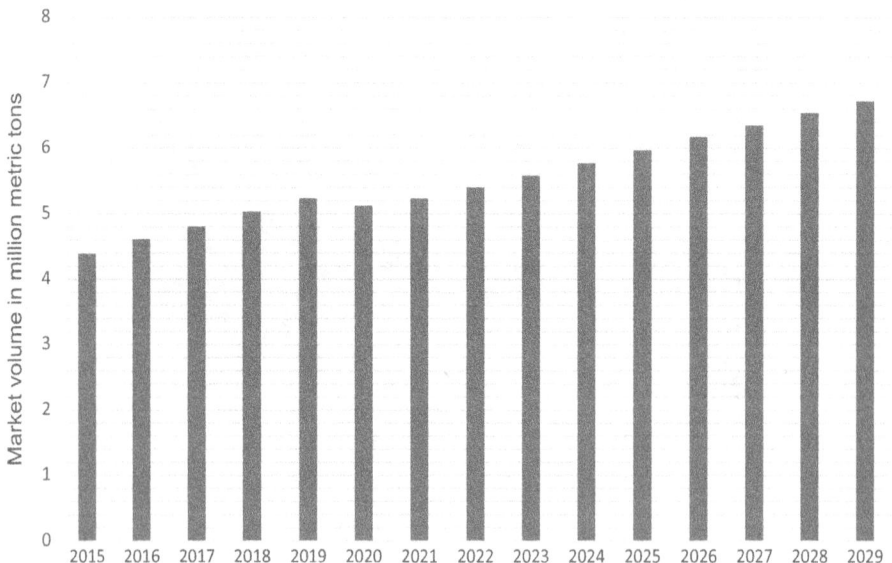

Figure 3.3: Market volume *n*-butanol around the world with a forecast of 2029.

performance decreases, and the amount of combustion wave is enhanced. Other alcohols can be used for this purpose such as propanol and alcohols with higher carbon contents [60–62].

3.4 Other oxygenates

Although other groups of oxygenates are not used as typical as ethers and alcohols, some of them have been promising in research.

3.4.1 Carbonates

Organic carbonates are environmentally friendly chemicals with several uses. They are frequently utilized in the production of essential commercial and industrial products. The most significant linear carbonates are dimethyl carbonate (DMC) and diethyl carbonate (DEC).

DMC is not expensive to produce and process. Its production needs methanol and ethylene carbonate, which are affordable. When trans-esterified, ethylene glycol is also created along with DMC, which is profoundly valuable for industry. DMC has the most effect when blended with gasoline or diesel at 5–10% regarding thermal efficiency, better combustion, and reduced NO_x emission [2, 63, 64].

DEC has not been studied as profoundly as DMC. However, it has been proven that DEC as an additive can reduce soot and smoke emissions [65].

3.4.2 Esters

Ester additives outperform ethanol and ethers in various ways. They are not toxic, do not produce carbon monoxide, and are cheaper to produce. Here, we will discuss methyl acetate and ethyl acetate briefly [66].

Methyl acetate ($C_3H_6O_2$) is a carboxylate ester and a flammable liquid that is produced by two-step methylation of acetic acid. It has been promising in decreasing CO emissions in diesel engines while increasing NO_x [67].

Another compound, ethyl acetate, has been pointed out to generate a more stable flame than ethanol or gasoline in the engine, increase water tolerance of ethanol–gasoline blend and octane number, and decrease organic emissions [68, 69].

Both methyl acetate and ethyl acetate increase fuel's octane rating by at least three folds [66].

References

[1] Arteconi A, Mazzarini A, di Nicola G. Emissions from ethers and organic carbonate fuel additives: A review. *Water, Air, and Soil Pollution* 2011;**221**: 405–23. doi:10.1007/S11270-011-0804-Y

[2] Gopinath D, Sundaram EG. Experimental investigation on the effect of adding di methyl carbonate to gasoline in a SI engine performance. *International Journal of Scientific & Engineering Research* 2012;**3**: 1–5.

[3] Anderson JE, Kramer U, Mueller SA, *et al*. Octane numbers of ethanol- and methanol-gasoline blends estimated from molar concentrations. *Energy and Fuels* 2010;**24**: 6576–85. doi:10.1021/ef101125c

[4] Badra J, Alowaid F, Alhussaini A, *et al*. Understanding of the octane response of gasoline/MTBE blends. *Fuel* 2022;**318**: 123647. doi:10.1016/j.fuel.2022.123647

[5] Cataluña R, Dalávia D, Da Silva R, *et al*. Acceleration tests using gasolines formulated with di-TAE, TAEE and MTBE ethers. *Fuel* 2011;**90**: 992–6. doi:10.1016/j.fuel.2010.10.031

[6] Patil AR, Taji SG. Effect of Oxygenated Fuel Additive on Diesel Engine Performance and Emission: A Review. *IOSR Journal of Mechanical and Civil Engineering*:30–5.www.iosrjournals.org (accessed 2 Jul 2022).

[7] Ershov MA, Potanin DA, Tarazanov S v., *et al*. Blending Characteristics of Isooctene, MTBE, and TAME as Gasoline Components. *Energy and Fuels* 2020;**34**: 2816–23. doi:10.1021/ACS.ENERGYFUELS.9B03914

[8] Leone TG, Anderson JE, Davis RS, *et al*. The Effect of Compression Ratio, Fuel Octane Rating, and Ethanol Content on Spark-Ignition Engine Efficiency. *Environmental Science and Technology* 2015;**49**: 10778–89. doi:10.1021/acs.est.5b01420

[9] Wei H, Yao C, Pan W, *et al*. Experimental investigations of the effects of pilot injection on combustion and gaseous emission characteristics of diesel/methanol dual fuel engine. *Fuel* 2017;**188**: 427–41. doi:10.1016/j.fuel.2016.10.056

[10] Kiatphuengporn S, Donphai W, Jantaratana P, *et al*. Cleaner production of methanol from carbon dioxide over copper and iron supported MCM-41 catalysts using innovative integrated magnetic field-packed bed reactor. *Journal of Cleaner Production* 2017;**142**: 1222–33. doi:10.1016/j.jclepro.2016.08.086

[11] Mishra PC, Kar SK, Mishra H. Effect of perforation on exhaust performance of a turbo pipe type muffler using methanol and gasoline blended fuel: A step to NOx control. *Journal of Cleaner Production* 2018;**183**: 869–79. doi:10.1016/j.jclepro.2018.02.236

[12] Olah GA. Beyond oil and gas: The methanol economy. *Angewandte Chemie – International Edition* 2005;**44**: 2636–9. doi:10.1002/anie.200462121

[13] Biernacki P, Röther T, Paul W, *et al*. Environmental impact of the excess electricity conversion into methanol. *Journal of Cleaner Production* 2018;**191**: 87–98. doi:10.1016/j.jclepro.2018.04.232

[14] Thakur AK, Kaviti AK, Mehra R, *et al*. Performance analysis of ethanol–gasoline blends on a spark ignition engine: a review. *Biofuels* 2017;**8**: 91–112. doi:10.1080/17597269.2016.1204586

[15] Amirabedi M, Jafarmadar S, Khalilarya S. Experimental investigation the effect of Mn2O3 nanoparticle on the performance and emission of SI gasoline fueled with mixture of ethanol and gasoline. *Applied Thermal Engineering* 2019;**149**: 512–9. doi:10.1016/j.applthermaleng.2018.12.058

[16] Badra J, AlRamadan AS, Sarathy SM. Optimization of the octane response of gasoline/ethanol blends. *Applied Energy* 2017;**203**: 778–93. doi:10.1016/j.apenergy.2017.06.084

[17]   İlhak Mİ, Doğan R, Akansu SO, *et al.* Experimental study on an SI engine fueled by gasoline, ethanol and acetylene at partial loads. *Fuel* 2020;**261**. doi:10.1016/j.fuel.2019.116148

[18]   Al-Hasan M. Effect of ethanol-unleaded gasoline blends on engine performance and exhaust emission. *Energy Conversion and Management* 2003;**44**: 1547–61. doi:10.1016/S0196-8904 (02)00166-8

[19]   Kassem MGA, Ahmed AMM, Abdel-Rahman HH, *et al.* Use of Span 80 and Tween 80 for blending gasoline and alcohol in spark ignition engines. *Energy Reports* 2019;**5**: 221–30. doi:10.1016/j.egyr.2019.01.009

[20]   Jin D, Choi K, Myung CL, *et al.* The impact of various ethanol-gasoline blends on particulates and unregulated gaseous emissions characteristics from a spark ignition direct injection (SIDI) passenger vehicle. *Fuel* 2017;**209**: 702–12. doi:10.1016/j.fuel.2017.08.063

[21]   Lim CS, Lim JH, Cha JS, *et al.* Comparative effects of oxygenates-gasoline blended fuels on the exhaust emissions in gasoline-powered vehicles. *Journal of Environmental Management* 2019;**239**: 103–13. doi:10.1016/j.jenvman.2019.03.039

[22]   Iodice P, Senatore A. Influence of Ethanol-gasoline Blended Fuels on Cold Start Emissions of a Four-stroke Motorcycle. Methodology and Results. 2013. doi:10.4271/2013-24-0117

[23]   Kareddula VK, Puli RK. Influence of plastic oil with ethanol gasoline blending on multi cylinder spark ignition engine. *Alexandria Engineering Journal* 2018;**57**: 2585–9. doi:10.1016/ j.aej.2017.07.015

[24]   Geng P, Zhang H. Combustion and emission characteristics of a direct-injection gasoline engine using the MMT fuel additive gasoline. *Fuel* 2015;**144**: 380–7. doi:10.1016/j. fuel.2014.12.064

[25]   Elfasakhany A. Investigations on the effects of ethanol–methanol–gasoline blends in a spark-ignition engine: Performance and emissions analysis. *Engineering Science and Technology, an International Journal* 2015;**18**: 713–9. doi:10.1016/j.jestch.2015.05.003

[26]   Yilmaz İ, Taştan M. Investigation of hydrogen addition to methanol-gasoline blends in an SI engine. *International Journal of Hydrogen Energy* 2018;**43**: 20252–61. doi:10.1016/ j.ijhydene.2018.07.088

[27]   Production capacity of methanol worldwide from 2018 to 2020, with a forecast for 2030 (in million metric tons). 2021. https://www.statista.com/statistics/1065891/global-methanol -production-capacity/%09%09%09%09%09%09%09%09%09%09%0A

[28]   Riaz A, Zahedi G, Klemeš JJ. A review of cleaner production methods for the manufacture of methanol. *Journal of Cleaner Production* 2013;**57**: 19–37. doi:10.1016/j.jclepro.2013.06.017

[29]   Chen M, Schmidt LD. Morphology and composition of PtPd alloy crystallites on SiO2 in reactive atmospheres. *Journal of Catalysis* 1979;**56**: 198–218. doi:10.1016/0021-9517(79) 90107-6

[30]   Wan Z, Wu W, Li G (Kevin), *et al.* Effect of SiO2/Al2O3 ratio on the performance of nanocrystal ZSM-5 zeolite catalysts in methanol to gasoline conversion. *Applied Catalysis A: General* 2016;**523**: 312–20. doi:10.1016/j.apcata.2016.05.032

[31]   Firoozi M, Baghalha M, Asadi M. The effect of micro and nano particle sizes of H-ZSM-5 on the selectivity of MTP reaction. *Catalysis Communications* 2009;**10**: 1582–5. doi:10.1016/ j.catcom.2009.04.021

[32]   https://www.statista.com/statistics/1106345/distribution-of-global-ethanol-production-by- country/. 2021.

[33]   Fayyazbakhsh A, Pirouzfar V. Investigating the influence of additives-fuel on diesel engine performance and emissions: Analytical modeling and experimental validation. *Fuel* 2016;**171**: 167–77. doi:10.1016/j.fuel.2015.12.028

[34] Fayyazbakhsh A, Pirouzfar V. Determining the optimum conditions for modified diesel fuel combustion considering its emission, properties and engine performance. *Energy Conversion and Management* 2016;**113**: 209–19. doi:10.1016/j.enconman.2016.01.058

[35] Fayyazbakhsh A, Pirouzfar V. Comprehensive overview on diesel additives to reduce emissions, enhance fuel properties and improve engine performance. *Renewable and Sustainable Energy Reviews* 2017;**74**: 891–901. doi:10.1016/j.rser.2017.03.046

[36] Işik MZ, Aydin H. Investigation on the effects of gasoline reactivity controlled compression ignition application in a diesel generator in high loads using safflower biodiesel blends. *Renewable Energy* 2019;**133**: 177–89. doi:10.1016/j.renene.2018.10.025

[37] Ratcliff MA, Windom B, Fioroni GM, *et al.* Impact of ethanol blending into gasoline on aromatic compound evaporation and particle emissions from a gasoline direct injection engine. *Applied Energy* 2019;**250**: 1618–31. doi:10.1016/j.apenergy.2019.05.030

[38] Murata K, Inaba M, Takahara I. Effects of surface modification of H-ZSM-5 catalysts on direct transformation of ethanol into lower olefins. *Journal of the Japan Petroleum Institute* 2008;**51**: 234–9. doi:10.1627/jpi.51.234

[39] Ni Y, Peng W, Sun A, *et al.* High selective and stable performance of catalytic aromatization of alcohols and ethers over La/Zn/HZSM-5 catalysts. *Journal of Industrial and Engineering Chemistry* 2010;**16**: 503–5. doi:10.1016/j.jiec.2010.03.011

[40] Gayubo AG, Alonso A, Valle B, *et al.* Hydrothermal stability of HZSM-5 catalysts modified with Ni for the transformation of bioethanol into hydrocarbons. *Fuel* 2010;**89**: 3365–72. doi:10.1016/j.fuel.2010.03.002

[41] Hu Z, Zhang H, Wang L, *et al.* Highly stable boron-modified hierarchical nanocrystalline ZSM-5 zeolite for the methanol to propylene reaction. *Catal Sci Technol* 2014;**4**: 2891–5. doi:10.1039/C4CY00376D

[42] Lee J, Hong UG, Hwang S, *et al.* Production of light olefins through catalytic cracking of C5 raffinate over carbon-templated ZSM-5. *Fuel Processing Technology* 2013;**108**: 25–30. doi:10.1016/j.fuproc.2012.03.005

[43] Viswanadham N, Saxena SK, Kumar J, *et al.* Catalytic performance of nano crystalline H-ZSM-5 in ethanol to gasoline (ETG) reaction. *Fuel* 2012;**95**: 298–304. doi:10.1016/j.fuel.2011.08.058

[44] Market volume of n-Butanol worldwide from 2015 to 2020, with a forecast for 2021 to 2026. 2021. https://www.statista.com/statistics/1245211/n-butanol-market-volume-worldwide/

[45] Saini S, Chandel AK, Sharma KK. Past practices and current trends in the recovery and purification of first generation ethanol: A learning curve for lignocellulosic ethanol. *Journal of Cleaner Production* 2020;**268**: 122357. doi:10.1016/j.jclepro.2020.122357

[46] Amid S, Aghbashlo M, Tabatabaei M, *et al.* Effects of waste-derived ethylene glycol diacetate as a novel oxygenated additive on performance and emission characteristics of a diesel engine fueled with diesel/biodiesel blends. *Energy Conversion and Management* 2020;**203**. doi:10.1016/j.enconman.2019.112245

[47] Qureshi N, Saha BC, Cotta MA, *et al.* An economic evaluation of biological conversion of wheat straw to butanol: A biofuel. *Energy Conversion and Management* 2013;**65**: 456–62. doi:10.1016/j.enconman.2012.09.015

[48] Nguyen DD, Moghaddam H, Pirouzfar V, *et al.* Improving the gasoline properties by blending butanol-Al2O3 to optimize the engine performance and reduce air pollution. *Energy* 2021;**218**: 119442. doi:10.1016/j.energy.2020.119442

[49] Jin C, Yao M, Liu H, *et al.* Progress in the production and application of n-butanol as a biofuel. *Renewable and Sustainable Energy Reviews* 2011;**15**: 4080–106. doi:10.1016/j.rser.2011.06.001

[50] Feng R, Yang J, Zhang D, *et al.* Experimental study on SI engine fuelled with butanol-gasoline blend and H2O addition. *Energy Conversion and Management* 2013;**74**: 192–200. doi:10.1016/j.enconman.2013.05.021

[51] Elfasakhany A. Experimental investigation on SI engine using gasoline and a hybrid iso-butanol/gasoline fuel. *Energy Conversion and Management* 2015;**95**: 398–405. doi:10.1016/j.enconman.2015.02.022

[52] Pandey RK, Rehman A, Sarviya RM. Impact of alternative fuel properties on fuel spray behavior and atomization. *Renewable and Sustainable Energy Reviews* 2012;**16**: 1762–78. doi:10.1016/j.rser.2011.11.010

[53] Kalam MA, Masjuki HH. Biodiesel from palmoil – An analysis of its properties and potential. *Biomass and Bioenergy* 2002;**23**: 471–9. doi:10.1016/S0961-9534(02)00085-5

[54] Wei H, Feng D, Pan M, *et al.* Experimental investigation on the knocking combustion characteristics of n-butanol gasoline blends in a DISI engine. *Applied Energy* 2016;**175**: 346–55. doi:10.1016/j.apenergy.2016.05.029

[55] Knothe G, Steidley KR. Kinematic viscosity of biodiesel fuel components and related compounds. Influence of compound structure and comparison to petrodiesel fuel components. *Fuel* 2005;**84**: 1059–65. doi:10.1016/j.fuel.2005.01.016

[56] Zaharin MSM, Abdullah NR, Masjuki HH, *et al.* Evaluation on physicochemical properties of iso-butanol additives in ethanol-gasoline blend on performance and emission characteristics of a spark-ignition engine. *Applied Thermal Engineering* 2018;**144**: 960–71. doi:10.1016/j.applthermaleng.2018.08.057

[57] Szwaja S, Naber JD. Combustion of n-butanol in a spark-ignition IC engine. *Fuel* 2010;**89**: 1573–82. doi:10.1016/j.fuel.2009.08.043

[58] Wigg B, Coverdill R, Lee C-F, *et al.* Emissions Characteristics of Neat Butanol Fuel Using a Port Fuel-Injected, Spark-Ignition Engine. 2011. doi:10.4271/2011-01-0902

[59] Pechout M, Mazac M, Vojtisek-Lom M. Effect of Higher Content N-Butanol Blends on Combustion, Exhaust Emissions and Catalyst Performance of an Unmodified SI Vehicle Engine. 2012. doi:10.4271/2012-01-1594

[60] Karimi N, Pirouzfar V, Su C. Enhancing engine power and torque and reducing exhaust emissions of blended fuels derived from gasoline-propanol-nano particles. *Energy* 2022;**241**: 122924. doi:10.1016/j.energy.2021.122924

[61] Mourad M, Mahmoud KRM. Performance investigation of passenger vehicle fueled by propanol / gasoline blend according to a city driving cycle. *Energy* 2018;**149**: 741–9. doi:10.1016/j.energy.2018.02.099

[62] Alonso C, Montero EA, Chamorro CR, *et al.* Vapor – liquid equilibrium of octane-enhancing additives in gasolines 5. Total pressure data and G E for binary and ternary mixtures containing 1, 1-dimethylpropyl methyl ether (TAME), 1-propanol and n-hexane at 313. 15 K &. 2003;**212**: 81–95. doi:10.1016/S0378-3812(03)00268-1

[63] Patil AR, Bindu RS, Pawar AM, *et al.* Experimental investigation on the effect of optimized dimethyl carbonate on CI engine performance & Emissions at various engine operating parameters using Taguchi method. *AIP Conference Proceedings* 2022;**2469**. doi:10.1063/5.0080194

[64] Shukla K, Srivastava VC. Synthesis of organic carbonates from alcoholysis of urea: A review. *Catalysis Reviews – Science and Engineering* 2017;**59**: 1–43. doi:10.1080/01614940.2016.1263088

[65] Aguado-Deblas L, Hidalgo-Carrillo J, Bautista FM, *et al.* Biofuels from diethyl carbonate and vegetable oils for use in triple blends with diesel fuel: Effect on performance and smoke emissions of a diesel engine. *Energies (Basel)* 2020;**13**. doi:10.3390/EN13246584

[66] Dabbagh HA, Ghobadi F, Ehsani MR, *et al.* The influence of ester additives on the properties of gasoline. *Fuel* 2013;**104**: 216–23. doi:10.1016/J.FUEL.2012.09.056

[67] Londhe H, Luo G, Park S, *et al.* Testing of anisole and methyl acetate as additives to diesel and biodiesel fuels in a compression ignition engine. *Fuel* 2019;**246**: 79–92. doi:10.1016/J.FUEL.2019.02.079

[68] Badawy T, Williamson J, Xu H. Laminar burning characteristics of ethyl propionate, ethyl butyrate, ethyl acetate, gasoline and ethanol fuels. *Fuel* 2016;**183**: 627–40. doi:10.1016/J.FUEL.2016.06.087

[69] Amine M, Awad EN, Ibrahim V, *et al.* Influence of ethyl acetate addition on phase stability and fuel characteristics of hydrous ethanol-gasoline blends. *Egyptian Journal of Petroleum* 2018;**27**: 1333–6. doi:10.1016/J.EJPE.2018.09.005

4 Previous research on the influence of oxygenated additives blended with gasoline on gasoline factors

4.1 Engine and combustion performance

The influence of different types of alcohols on engine performance factors is considered and given below.

4.1.1 Brake-specific fuel consumption

Brake-specific fuel consumption (BSFC) is a factor for showing the fuel efficiency of any mover that burns fuel. It compares the efficiency of internal combustion (IC) engines. The power produced is the main factor in measuring BSFC. The BSFC decreases with increasing engine load in the gasoline-powered motor. This is because the air/fuel combination tends to be on the leaner side; therefore, stoichiometric ratios are used [1].

BSFC is changed by the oxygen content of the oxygenated additive. Because other fuel samples have a low heat value, the BSFC increases with the addition of ethanol concentration to maintain engine power [2]. Blending oxygenate additives can improve this factor because of their potential to complete the combustion by enhancing the fuel's oxygen content. Amirabedi et al. [3] reviewed the effect of ethanol and manganese oxide nano-metal blended with diesel on engine performance. Their tests included three parts, and the results for BSFC presented that this factor is enhanced via blending 10% of ethanol due to the higher vaporization latent heat of ethanol.

On the other hand, when 10 ppm was added to gasoline 10% ethanol, the results revealed a significant reduction in BSFC (approximately −35%). Methanol usually improves the BSFC of gasoline and sometimes shows better performance than ethanol owing to higher oxygen content; however, the main drawback of this additive is its limitation as it is poisonous. In some cases, 20–40% improvement in BSFC has been reported by researchers after blending methanol with gasoline which enhances the amount of fuel needed to get the same power as pure gasoline. Sarikoc [4] reported that blending 20% methanol with gasoline could improve the BSFC by about 20–30% in different engine loads. However, adding hydrogen reduced methanol's influence on BSFC, which has been reported by another researcher [5]. This could be due to the lower methanol calorific value and higher calorific value of hydrogen than gasoline [6]. Figure 4.1 compares the alcohol effects on gasoline with diesel engines regarding BSFC.

https://doi.org/10.1515/9783110999969-004

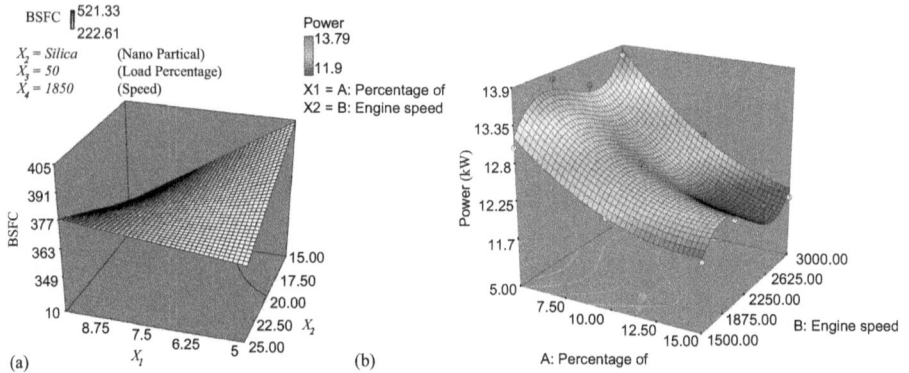

Figure 4.1: The difference between the influence of alcohols on diesel (a) and gasoline (b) BSFC.

4.1.2 Brake thermal efficiency and power

BTE is a critical factor in choosing an oxygenated or nano-particle additive. Blending oxygenated additives cause an increase in the BTE due to the higher latent vaporization heat. He et al. [7] investigated the effect of adding normal butanol to gasoline on combustion characteristics in the in-cylinder direct injection (DI) engine and the port injection (PI). They concluded that the combustion time was shortened, engine ignition time was advanced, and BTE was improved only by adding n-butanol. Yusoff et al. [8] suggested that the brake power is improved in an almost linear fashion when the engine speed increases. Moreover, they concluded that most blends have no considerable result of brake power. The normal-Bu20 and iso-Bu20 blend have higher brake power, whereas gasoline reflects a negligible increment of 1.36 and 1.73%, respectively. Since iso-butanol and normal-butanol are essentially oxygenated when blended, the octane rating is increased. They may not be prone to auto-ignition, but they can improve the brake power of blended fuel. BTE is increased by fueling hydrous ethanol-gasoline instead of pure gasoline at the same tested engine speed. The main reason for such influence is associated with the fact that ethanol produces the hydroxyl radical (−OH) and leads the combustion to be more complete and improves flame propagation speed. Therefore, lower heat loss to the cylinder walls and shorter combustion periods are obtained, and BTE is enhanced. Sebayang et al. [9] suggested that the BTE of bioethanol-gasoline is higher than neat gasoline due to the high mean effective pressure of the blended fuel. Oh et al. [10] studied the influence of adding hydrogen nano-bubble (HNB) additive to gasoline on engine performance. Their results indicated a direct dependency between BTE and engine load because of the increase in the air-fuel rate. Also, they reported that BTE in the HNB gasoline combustion is higher than neat gasoline fuel at all engine loads.

Table 4.1 summarizes some of the studies' key findings considering the effects of different alcohols on gasoline engines.

Table 4.1: The influence of different alcohols on the engine performance of the gasoline-powered engine.

Engine	Power property	Additive	Additive content	Influence on the property	Main findings
Four-stroke multi-cylinder MPFI SI engine coupled with an eddy current dynamometer	BSFC BTE	Ethanol	6.25% 10% 20% 6.25% 10% 20%	≈−1.25 ≈−3.5% ≈−6.3% ≈2% ≈3% ≈4%	Higher engine speed means less time for fuel burning in the combustion chamber and causes lower BTE compared with lower engine speed. The influence of engine load is the opposite. And with ethanol, enhancing the engine speed causes incomplete combustion [2].
Four-cylinder, water-cooled automotive spark-ignition engine	BSFC	n-Butanol	2.50% 5% 7.50%	≈−7% ≈−6.3% ≈−5.3%	n-Butanol has a greater boiling temperature and heat of vaporization than gasoline. As a result, n-butanol is less likely to totally evaporate than gasoline. Particularly, high n-butanol ratios have a harmful influence on BSFC [11].
A 4-cylinder, inline, port fuel injection, high-speed SI engine	Power BSFC	n-Butanol	20% 30% 20% 30%	≈3.2% ≈5.2% ≈−2.5% ≈−3.1%	The combustion process is significantly different in fuel-lean and fuel-rich blends. So, BSFC reduces via enhancing butanol when the engine works at fuel-lean and vice versa [12].
Inline four-cylinder water-cooled gasoline engine	BSFC	Butanol Butanol + EGR + EGR	10% 20% 10% B + 6% E 20% B + 6% E 12% 18%	≈−1% ≈−1.5% ≈−3% ≈−4% ≈−7% ≈−8.5%	A higher exhaust gas recirculation (EGR) ratio can improve throttle opening that reduces throttle and pumping losses and raise the engine efficiency [13].

Table 4.1 (continued)

Engine	Power property	Additive	Additive content	Influence on the property	Main findings
1.8 L turbocharged GDI engine	BTE	EGR EGR + air	10% 20% 10% EGR 20% EGR	≈4% ≈8% ≈8% ≈15%	Because the flame propagation speed reduces due to EGR, the burning duration and engine power increase. This influence is stronger with air due to a reduction in the burning temperature [14].

4.2 Exhaust emissions

4.2.1 Greenhouse gases

The greenhouse effect is a natural process that leads to global warming. Methane, fluorinated gases, CO_2, nitrous oxide (N_2O), and water vapor are mainly responsible for this effect. Among all greenhouse gases, water vapor is responsible for most greenhouse effects [15]. Unlike other GHGs, the concentration of water vapor cannot vary. However, the temperature is the only factor that can make the concentration of water vapor different in different places. Surprisingly, despite most people who believe that CO_2 is the most significant characteristic of this effect, the vapor is responsible for almost 60% of this effect. It is worth mentioning that its concentration remains stable in the atmosphere in comparison with other GHGs.

Moreover, it is a factor to enhance the amount of CO_2 in the atmosphere. Due to the latest Intergovernmental Panel on Climate Change (IPCC) report, the amount of CO_2 will increase till 2100 due to most scenarios [16]. Luckily, the regulation made in Paris made the countries more responsible for their actions and the number of emissions they are creating, especially CO_2 [17]. Although a vast thrive is inducing to avoid this emission, researchers believe that to stabilize the world temperature; there should be more reduction in CO_2 [18]. Moreover, it has been proven that the best technique to reduce this emission is stopping using the majority of types of energy, especially gasoline and diesel. In this case, in the COVID-19 pandemic, due to restrictions in the developed countries, the amount of CO_2 reduced sharply in 2019 and 2020 compared to before that time [19]. Figure 4.2 shows the distribution of GHG emissions by sector in 2016 [20]. As can be seen, road transportation has the most considerable influence on this effect in which gasoline-powered engines are the central part.

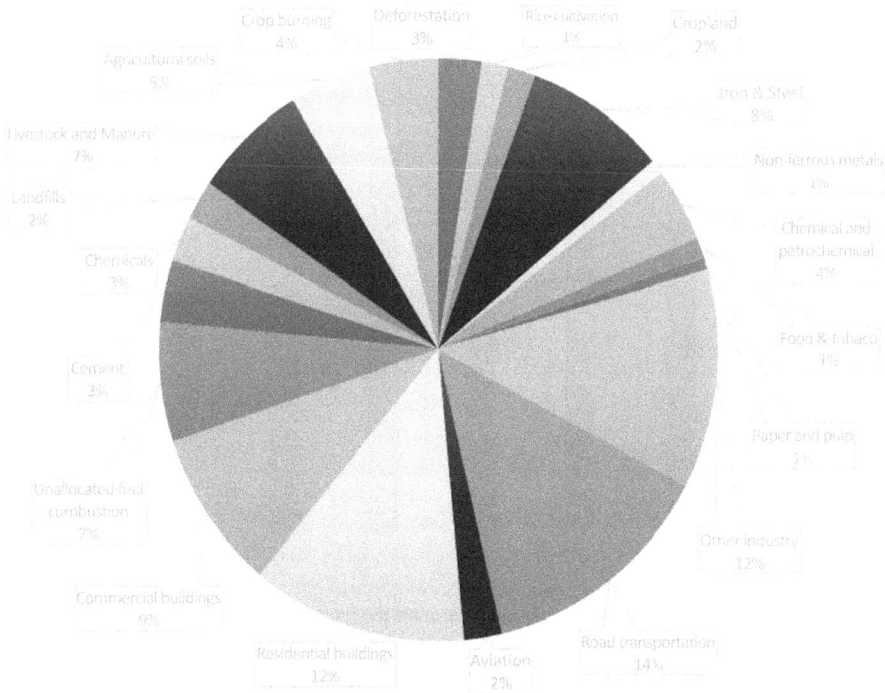

Figure 4.2: Distribution of greenhouse gas emissions around the world by sector in 2016.

4.2.2 Carbon monoxide

Much research has been done on carbon monoxide (CO) and CO_2 emissions due to their importance worldwide [21, 22]. CO is one of the main reasons for lung cancer in polluted cities. GHGs, especially CO_2, cause climatic changes. In this case, the earth's temperature would be −15 without GHGs [23]. Figure 4.3 shows and forecasts carbon dioxide levels from 2018 to 2050 [24].

Hasan et al. [25] used ethanol as an additive to improve engine performance and reduce air pollution. They found that ethanol content can reduce CO_2 emission while having dual effects on CO emission. When the compression ratio is higher, ethanol declines CO emissions. On the contrary, it has an opposite effect at a lower compression ratio. CO and CO_2 emissions are affected by the engine's changing compression ratios and its operation. Sakthivel et al. [23] suggested that CO and HC emissions are significantly reduced at all vehicle speeds with higher alcohol content because of complete combustion. Akansu et al. [26] reported that CO emission of fuel with 20% ethanol is higher at low engine loads than neat gasoline. Owing to the incomplete combustion, CO emission when using E20 is high. High thermal efficiency diminishes the CO emission when adding E20 at a high engine load.

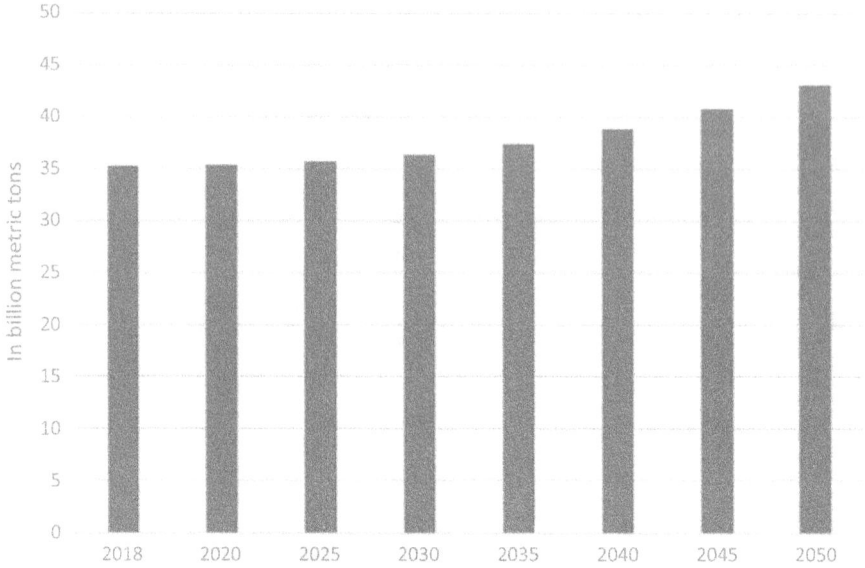

Figure 4.3: Forecast of carbon dioxide emissions around the world.

Additionally, they suggested that hydrogen blending with this fuel decreases the CO emission because its chemical content is zero carbon. Oh et al. [10] concluded that adding hydrogen can improve internal combustion engines [27, 28]. Furthermore, hydrogen contains potential energy to reach CO, CO_2, and HC to their lowest amount [29]. Later, Yilmaz et al. [30] used the combination of hydrogen and methanol. In this method, hydrogen can be served both as a fuel and additive. It is also derived from the synthesis of resources like water, fossil fuels, and biomass. A summary of prominent research main findings is listed in Table 4.2. Additionally, Figure 4.4 shows the influence of ethanol on CO emissions.

4.2.3 Nitrogen oxides

Nitrogen oxides (NOx) are mainly made of NOx and nitrogen dioxide. While the air dust is full of NOx, the generation of NO_X depends on the ignition delay and the engine temperature [9]. The following reactions affect NOx creation according to the mechanism of Zeldovich [37]:

$$N_2 + O \rightarrow NO + N$$

$$N + O_2 \rightarrow NO + O$$

$$N + OH \rightarrow NO + H$$

Table 4.2: The influence of different additives on the CO emissions of gasoline-powered engines.

Engine	Additive	Content	Influence on CO	Main outcomes
Single-cylinder, four-stroke, water-cooled, variable compression ratio, and PFI engine	Ethanol Lemon peel oil (LPO) Ethanol + LPO	20% 40% 20%E + 40% LPO	≈−33 ≈−32% ≈−63%	Biofuel blends have reduced CO emissions because they include fuel oxygen, which actively participates in the combustion to increase CO oxidation. Higher content of ethanol means higher oxygen content in the combustion process and lower CO emission [31].
In-line four-cylinder water-cooled gasoline engine	EGR EGR Ethanol EGR + ethanol	12% 24% 15% EGR 24% + Eth30%	≈−50% ≈+170% ≈+40% ≈+50%	At low EGR, CO decrease as a modest quantity of exhaust gas injection will not noticeably interfere with the mixture's combustion. In contrast, at high EGR, CO increases as the mixture cannot burn easily and completely [13].
Four-cylinder, four-stroke, spark-ignition, water-cooled	bioethanol Ag/Nb co-doped TiO$_2$ nanoparticle (TNA) bioethanol + TNA	20% 100 ppm B20% + TNA 100 ppm	≈−33% ≈−52% ≈−60%	Titanium, silver, and niobium metal elements operate as the high-energy third body, breaking chemical chains and producing unstable radicals that lead to improved combustion in the cylinder [32].
SI engine with combined injection with 1.984 L working volume	Anhydrous ethanol Hydrous ethanol Hydrous ethanol	– with 20% water dual fuel With 50% water dual fuel	≈+8% ≈−12% ≈−48%	In terms of physical effect, increasing water content resulted in a greater region of over-concentrated mixture in the cylinder and concentration of oxygen reduced by water oxygen, lowering the flame temperature and enhancing CO emissions [33].

Table 4.2 (continued)

Engine	Additive	Content	Influence on CO	Main outcomes
Single-cylinder spark-ignition	Ethanol (at vehicle speed 50 km/h)	10% 20% 50%	≈−15% ≈−24% ≈−51%	An increase in ethanol enhances the oxygen content of the fuel required for complete combustion. It can also reduce CO emissions as it is responsible for converting CO to CO_2 [34].
Single-cylinder and four-stroke gasoline engines	Fusel oil at load 2000 Fusel oil at load 4000	15% 30% 15% 30%	≈−20% ≈−34% ≈−17% ≈−40%	Enhance in fusel oil content cause enough time and oxygen for the combustion in the cylinder, two main factors for the eradication of CO emission [35].
In-line, four-cylinder, electronic fuel injection, automotive spark-ignition engine	Hydrous ethanol (2,000 rpm engine speed) Hydrous ethanol (2,500 rpm engine speed)	10% 20% 10% 20%	≈−26% ≈−31% ≈−27.5% ≈−43%	CO to CO_2 increased by ethanol. Moreover because of its higher oxygen content, ethanol made the mixed gas more homogeneous and caused complete combustion to trigger lower CO emissions [36].

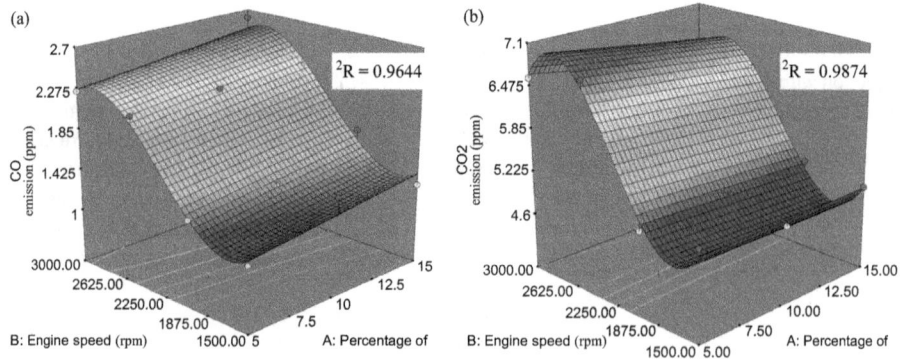

Figure 4.4: The influence of different variables and ethanol on CO emissions.

Streams of reactions including NO lastly produce nitric acid compounds as shown in Figure 4.5.

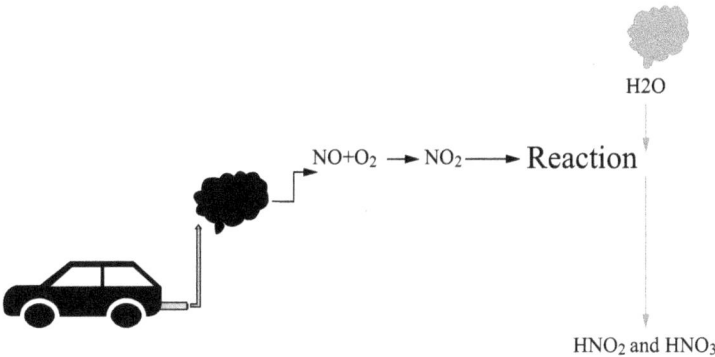

$$NO+O_2 \rightarrow NO_2 \longrightarrow \text{Reaction}$$

H2O

HNO$_2$ and HNO$_3$

Figure 4.5: NOx volume in the United States 1970–2020.

Most previous research indicates that blending alcohol with gasoline causes the development of nitrogen oxides due to alcohol's latent heat of vaporization, which provides higher temperatures (1600°C or more) required for the reaction of nitrogen with oxygen [38]. The acceleration of NOx emissions has been featured in most results using gasoline–bioethanol-blended fuels compared with neat gasoline [39, 40]. Other research contradicts the previous studies, which claimed a decrease in NOx, CO, and improved engine performance [41]. Another study dealt with adding the additives mentioned above and reported that all emissions are reduced through blending bioethanol with gasoline: NOx reduces about 10–15%when blending the fuel with E20 to E60 [42]. Li et al. [43] premeditated the impact of using acetone–butanol–ethanol (ABE) as a gasoline additive on NOx emissions. They specified that a sample with 30% of ABE has slightly higher NOx emissions. Because ethanol's oxygen content and latent vaporization heat are higher than butanol, the influence of 30% butanol blended with gasoline includes less than 30% of ABE. Likewise, they discovered that higher NOx emissions are specified by enhancing engine load due to higher cylinder temperatures. Previous research [44, 45] reported that newer engines could better control NO$_X$ emissions when using ethanol down to its better-sophisticated engine control strategies and after-treatment devices. Figure 4.6 exhibits NOx emissions in the United States between 1970 and 2020 [46]

Manigandan et al. [47] concluded that different ratios of hydrogen and exhaust gas recirculation (EGR) and TiO$_2$ 5% decrease the emissions of NO$_X$, CO, and CO$_2$ due to the effect of hydrogenated fuel. Hydrogen reduces flame development, rapid combustion, and combustion duration while using EGR increases the above parameters. Vipin and Subramanian [48] reported that water injection efficiently affects emission and knocking compared to EGR. Figure 4.7 presents NOx emissions in various conditions with butanol blending [18].

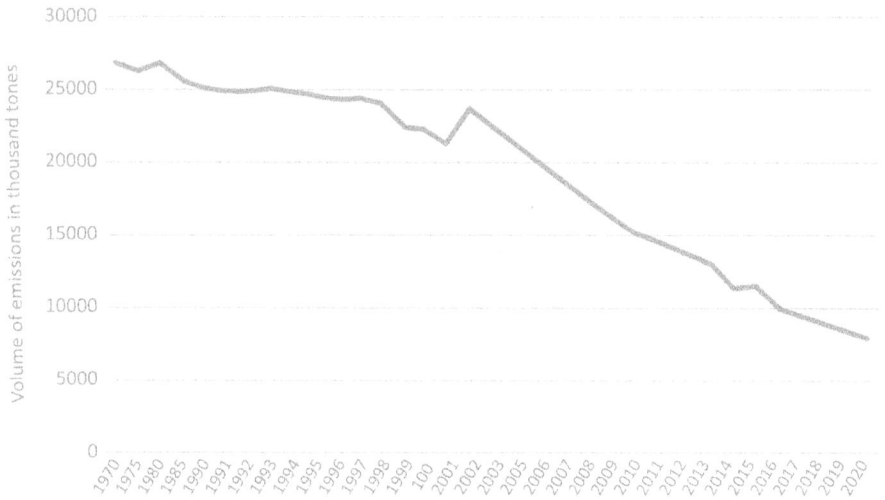

Figure 4.6: The mechanism of the NO reactions for generating nitric acid.

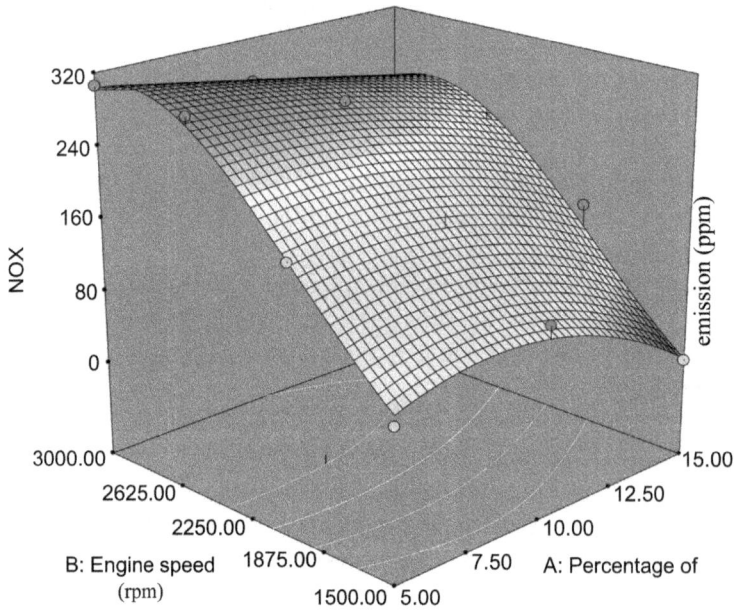

Figure 4.7: The results of NOx emissions in various engine speed conditions and butanol percentages in blended fuels.

4.2.4 Unburned hydrocarbons

Hydrocarbon (HC) is generally produced for the sake of incomplete combustion [49]. As the oxygenated additives can contribute to complete combustion, increasing the percentage of such additives causes a decline in the number of unburnt hydrocarbons. Elfasakhany [50] worked on the influence of blending methanol and ethanol with gasoline and its effect on air pollution at various engine speeds. He reported that two critical reasons are leading HC and CO to decrease. The first one is the high boiling point of gasoline. Since a higher boiling point contains components that may not be wholly burnt, it increases CO and UHC emissions. The second reason is that ethanol and methanol have higher latent heat of vaporization, enhancing CO_2 and reducing HC. In another research, Iodice et al. [51] considered the impact of ethanol blended with gasoline on HC emission. They reported that the complete combustion could be increased because of higher oxygen content and faster flame speed of blended fuel than pure gasoline. Hence, it is the main reason for lower HC emissions. One of the reasons for using nano-particles is to reduce these types of emissions. Wang et al. [52], who blended hydrous ethanol with gasoline, claimed that HC productions, caused by unburned materials, are located around the periphery's reaction regions. At the lower loads, the mixture of lower combustion temperature increases HC emission caused by flame quenching on the chamber walls. Increasing the percentage of ethanol (hydrous ethanol), the rate of oxygen causes a reduction in "lean outer flame," reducing HC emissions.

Additionally, the HC emissions bring about other products like particulate nucleation. It occurs at high exhaust gas temperatures [53]. Geng et al. [54] added methyl-cyclopentadienyl manganese tricarbonyl (MMT) to gasoline to detect the fuel emission. This study used a gasoline direct injection (GDI) engine with a speed of 2,000 rpm and the EURO V ultralow sulfur gasoline. Compared with other additives, the steam pressure of MMT is lower, and it is also less soluble [55]. Based on the results and tables, it is inferred that the HC emissions are remarkably diminished because it is homogenously blended, and the combustion efficiency is increased [56]. As mentioned earlier, nano-particles mainly reduce fuel emissions [57]. They may also cause converse outcomes. For instance, Valihesari et al. [58] analyzed the Fe_2O_3 and TiO_2 nano-particles as a gasoline additive in their research. They indicated that the flame silencing delay is increased while reaching the cylinder wall at the engine's highest speed for the high temperature of the combustion chamber. The TiO_2 nano-particles additive decreases the HC emissions, whereas the Fe_2O_3 nano-particles enhance the HC emission because incomplete combustion causes emissions at higher speeds.

As for HC emissions in different situations, see Figure 4.8 [59].

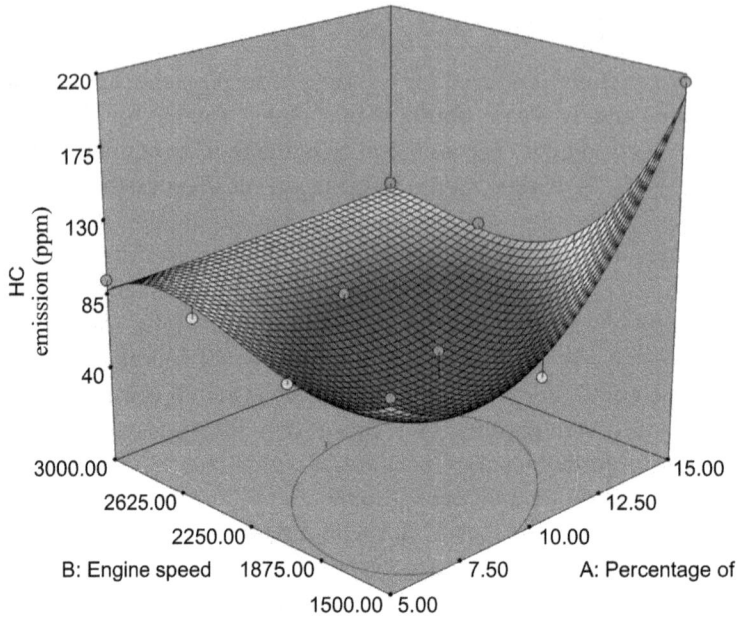

Figure 4.8: The results of HC emissions in various conditions of engine speed and butanol percentages in blended fuels.

4.2.5 Soot emission

The primary purpose of adding alcohol is to reduce soot emission, while it can be counted only as one reason for using nano-metals. However, nano-metals can also be utilized to boost thermal performance in fuel [37]. Polyaromatic compounds or Polycyclic aromatic hydrocarbons (PAH) are considered the most dangerous compound in soot due to their mutagenic, carcinogenic, and teratogenic toxicities as well as health problem such as blood cancer. Worldwide fatality rate because of particulate matters less than 25 μm is displayed in Figure 4.9 [60].

The blending of ethanol with gasoline has been experimented in engines [61–64] and flame temperature [65]. It was demonstrated that the blended fuels in various gasoline engines could reduce PAH and sharply soot emissions [66]. Considering most of the studies ever done on the influence of adding ethanol to gasoline, some researchers reported different results. They reported that this fuel could also reduce CO and NOx emissions [67, 68]; the emission of aldehydes [65] may also rise. Maricq et al. [69] analyzed the effect of adding different percentages of ethanol to gasoline at different fuel flow rates on soot formation. The amount of ethanol in the blend resulted in an undeniable reduction in soot formed and its size. Inal et al. [70] scrutinized methanol influences addition on the formation of PAHs and soot by light scattering extinction and

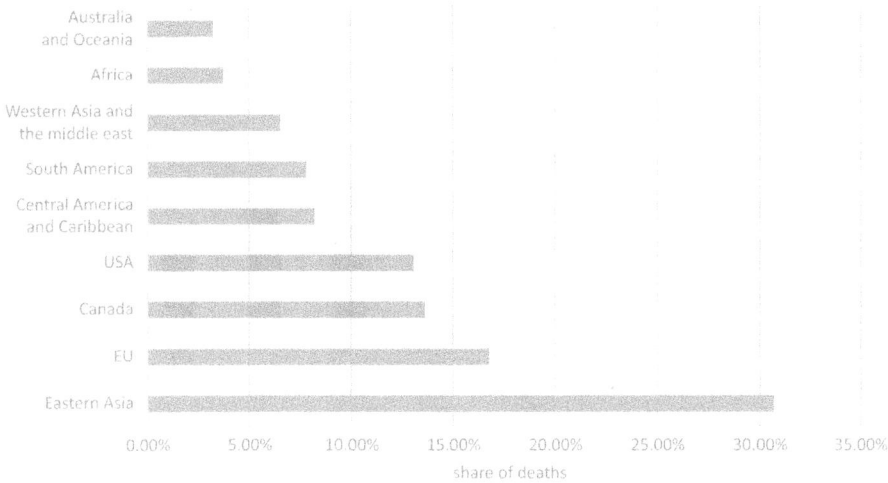

Figure 4.9: Regional deaths caused by PM2.5 emissions in 2018 worldwide.

gas chromatography-mass spectrometry techniques. The results indicated that the majority of detected aromatic was benzene. They determined that adding methanol could lessen PAH formation and soot following its oxygen content, directing into complete combustion. Li and Zou [71] focused on Ag and β-cyclodextrins, which were related to 5% of the nano-particles of TiO_2 to produce $CD\text{-}TiO_2\text{-}Ag\beta$.

On the other hand, limiting the quantity of sulfur and gasoline emissions is counted as an essential priority in the petroleum refining industry. Different methods sulfurize gasoline. Non-hydrodesulfurization is regarded as a fast and economical technology. This method contains catalytic hydro-desulfurization, fluid catalytic cracking (FCC), photocatalytic oxidation, oxidative biological, and adsorptive desulfurization [72, 73]. TiO_2 comprehensively wipes out environmental pollution. It is also noteworthy because it is cheap, nontoxic, and durable.

Furthermore, it has unique optical specifications, catalytic and photoelectric conversion potential, and properties. It is also a metallic oxide that acts as an absorbent in sulfurization processes [74, 75]. Wang et al. [76] deliberated the blending of ZSM-5 with ZrO_2, titanium, and NiO for sulfurization of FCC.

Considering the limits of HDS, other technologies such as bio-desulfurization [51, 77], selective adsorption [78], physical extraction [79], oxidative desulfurization (ODS) [80, 81] complexation, pervaporation [82] can be applied to eliminate the thiophene compounds. Among the items mentioned above, it is turned out that ODS is the most suitable way for gasoline sulfurization in FCC. It is a good option for a high percentage of sulfurization and not losing the octane number.

The European Union has seen a gradual decline in PM25 emissions since 2000. More detailed information is given in Figure 4.10 [83].

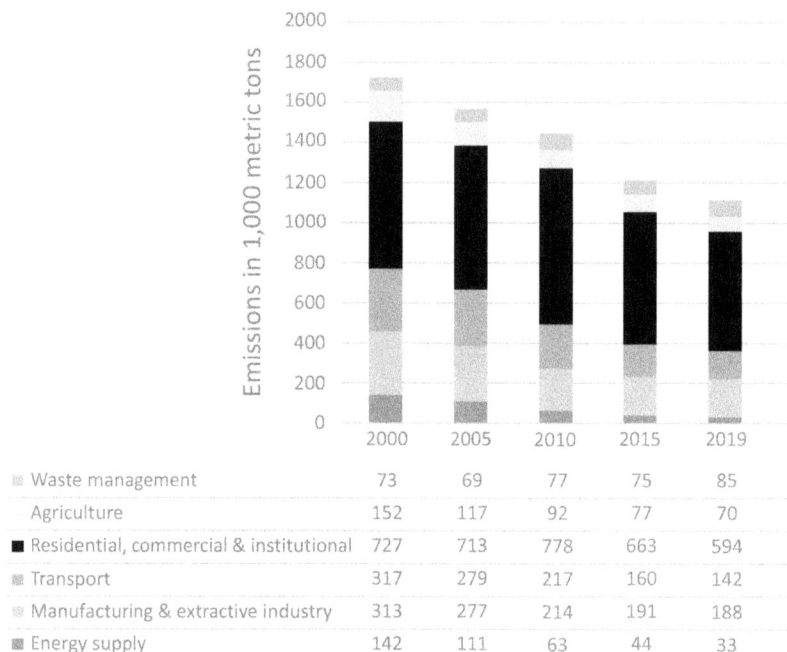

	2000	2005	2010	2015	2019
Waste management	73	69	77	75	85
Agriculture	152	117	92	77	70
Residential, commercial & institutional	727	713	778	663	594
Transport	317	279	217	160	142
Manufacturing & extractive industry	313	277	214	191	188
Energy supply	142	111	63	44	33

Figure 4.10: Total particulate matter emissions (PM2.5) in the European Union (EU-28) from 2000 to 2019, by sector.

4.2.6 Sulfuric emissions

Both natural and artificial processes are responsible for the sulfur dioxide emission (SO_2) into the atmosphere. Its emissions mainly rise from industrial activities such as petroleum refining and metal smelting, where fossil fuels are heavily consumed. We should be concerned about SO_2 emissions into the atmosphere because of the harm they do to the environment and human health on a local to a global scale, such as acid rains. Studies have pointed out that SO_2 emissions can travel very long distances. To illustrate, the ones originating from the Asian continent may reach the rest of the northern hemisphere, including North America [84–86].

When it comes to environmental effects, sulfur is a one-of-a-kind problem. It is difficult and expensive to be obliterated from petroleum, which naturally includes sulfur in different amounts. Only a tiny percentage of the contaminants remains after distillation and breaking, but they are still present in the gasoline. The catalyst in the catalytic converter will be poisoned if the sulfur content in the exhaust reaches a certain level. An exhaust system's catalyst loses its capacity to store oxygen if sulfur is allowed to flow through it, reducing the conversion process' efficiency.

Eliminating sulfur compounds improves combustion engines' lifetime and fosters positive attitudes towards the environment. For some years, researchers have focused their attention on clean fuel in an attempt to reduce pollution and halt the entry of contaminants into the atmosphere. As a result, several countries have enacted rigorous regulations to reduce the sulfur content in public transportation fuels like gasoline and diesel to less than 10 ppm [87–89].

Hydro-sulfurization is one of the traditional techniques used for sulfur removal. Although it eliminates many sulfuric compounds, this method is relatively expensive and complicated, especially for the removal of some content. Other alternative processes include oxidative, adsorptive, and bio-desulfurization. Accordingly, alumina, mesoporous silica, activated carbon, and porous graphitic carbon have been suggested to eliminate sulfuric material [87, 90–94].

Developed countries' efforts to reduce SO_2 emissions have resulted in enormous progress in lowering sulfuric pollution during the previous several decades. Regional and global ecosystems have significantly benefited as a consequence of these initiatives. A majority of developing countries have yet to execute effective strategies, while the minority has just recently started to do so. China, for example, didn't begin enforcing firm SO_2 pollution limits until far into the 2000s. Nonetheless, China has reduced the emissions significantly since 2013, when they issued a significant pollution regulation. On the other hand, SO_2 emissions in India and many other developing countries have continued to rise over the last few decades owing to the growth in energy consumption and the absence of efficient emission controls [86, 95–98].

4.3 Fuel properties

The influence of various types of additives on fuel properties is summarized as follows.

4.3.1 Octane number

The octane index is one of the main properties of spark-ignition fuel [99]. Octane number can be boosted by blending oxygenates additives such as ethanol, tertiary butyl alcohol (TBA), methyl tert-butyl ether (MTBE), methanol, normal-butanol, and tertiary butyl formate (TBF), and their blends with gasoline reduce the total cost. The gasoline octane index has a significant impact on the blending process of the fuel-air mixture as well as the combustion condition. Researchers have previously concluded that the proper octane index was required for achieving high efficiency and low emissions of GCI combustion [100]. Methanol has significant properties: (1) high octane rating even more than ethanol; (2) high vaporization latent heat; and (3) high oxygen percentage. In the oil industry, two mixing techniques apply alcohol as an octane improver: RON-mixing and splash-mixing blending techniques. The RON match-mixing technique is more complicated and usable. The splash-mixing technique is dependent on dosing the octane improver to the gasoline. The blended fuel would contain a higher-octane number in comparison with neat gasoline [101]. Morganti et al. [102] studied how blending methanol and ethanol improved the octane number. The results showed that methanol and ethanol enhance the octane number by more than 10%. Moreover, they claimed that methanol is more effective in improving the octane number.

4.3.2 Other properties

Studying gasoline's chemical and physical properties is essential as they are critical in analyzing and controlling air pollution and engine performance [103, 104]. Improving the chemical and physical properties such as viscosity and the latent heat of vaporization of gasoline are among the first priorities in blending after the octan number. In this case, properties such as viscosity or density are developed after blending alcohol with gasoline because alcohol influenced them. Different additives are utilized in this area. Diisopropyl ether (DIPE) is an additive used by Dhamodaran and his colleagues [105]. They suggested that blending DIPE with gasoline reduced the density and the calorific value in resultant blends because of its lower density of DIPE and its higher oxygen presence, respectively.

Furthermore, they reported that the blended fuel showed excellent volatility. Mixing DIPE with gasoline lessened the pressure of Reid vapor because DIPE does not offer any azeotropic behaviors. Rodriguez-Anton et al. [106] blended iso-butanol and ethyl tert-butyl ether (ETBE) to analyze the influence of iso-butanol and ETBE on the physical properties of gasoline. They stated that the lower oxygen percentage of iso-butanol and its higher density develop a higher energy density and a stoichiometric air per fuel ratio in the vicinity of gasoline in comparison with ethanol. These parameters make higher renewable energy for the same percentage of oxygen fuel. Although the energy density is lower than gasoline, the mixture's stoichiometric air/ fuel energy density is almost equal for all fuels. Henceforth, the maximum torque and power will be equal if the engine's fuel nozzles can supply the required fuel flow.

Additionally, they reported that iso-butanol has lower oxygen content than ethanol since it has a high potential due to its lower polarity effects, problems caused by corrosiveness, and the tendency of phase separation for fuel infrastructures and injection systems. Rodriguez et al. [107] contemplated the influence of ethanol-gasoline-ETBE fuel on Reid vapor pressure. They concluded that adding ETBE to gasoline reduces the Reid vapor pressure. This effect is directly related to ETBE percentage, while ethanol blending boosted the Reid vapor pressure of the mixing higher than 30%v/v. Therefore, increasing the ETBE content to compensate for the growth of Reid vapor pressure produced using ethanol at low concentrations. Nonetheless, the blended materials of both oxygenates could overcome the higher oxygen content limit.

References

[1] Qadiri U, AlFantazi A. Numerical 1-D simulations on Single-Cylinder stationary spark ignition engine using Micro-Emulsions, gasoline, and hydrogen in dual fuel mode. *Cleaner Chemical Engineering* 2022;2: 100009. doi:10.1016/j.clce.2022.100009

[2] Verma A, Dugala NS, Singh S. Experimental investigations on the performance of SI engine with Ethanol-Premium gasoline blends. *Materials Today: Proceedings* 2022;**48**: 1224–31. doi:10.1016/j.matpr.2021.08.255

[3] Amirabedi M, Jafarmadar S, Khalilarya S. Experimental investigation the effect of Mn2O3 nanoparticle on the performance and emission of SI gasoline fueled with mixture of ethanol and gasoline. *Applied Thermal Engineering* 2019;**149**: 512–9. doi:10.1016/j .applthermaleng.2018.12.058

[4] Sarıkoç S. Effect of H 2 addition to methanol-gasoline blend on an SI engine at various lambda values and engine loads: A case of performance, combustion, and emission characteristics. *Fuel* 2021;**297**: 120732. doi:10.1016/j.fuel.2021.120732

[5] Akar MA, Serin H, Tosun E. ScienceDirect Variation of spark plug type and spark gap with hydrogen and methanol added gasoline fuel: Performance characteristics. 2020;**5**. doi:10.1016/j.ijhydene.2020.03.110

[6] Duan X, Li Y, Liu J, *et al.* Experimental study the effects of various compression ratios and spark timing on performance and emission of a lean-burn heavy-duty spark ignition engine fueled with methane gas and hydrogen blends. *Energy* 2019;**169**: 558–71. doi:10.1016/j.energy.2018.12.029

[7] He BQ, Chen X, Lin CL, *et al.* Combustion characteristics of a gasoline engine with independent intake port injection and direct injection systems for n-butanol and gasoline. *Energy Conversion and Management* 2016;**124**: 556–65. doi:10.1016/ j.enconman.2016.07.053

[8] Yusoff MNAM, Zulkifli NWM, Masjuki HH, *et al.* Performance and emission characteristics of a spark ignition engine fuelled with butanol isomer-gasoline blends. *Transportation Research Part D: Transport and Environment* 2017;**57**: 23–38. doi:10.1016/j.trd.2017.09.004

[9] Sebayang AH, Masjuki HH, Ong HC, *et al.* Prediction of engine performance and emissions with Manihot glaziovii bioethanol – Gasoline blended using extreme learning machine. *Fuel* 2017;**210**: 914–21. doi:10.1016/j.fuel.2017.08.102

[10] Oh SH, Yoon SH, Song H, *et al.* Effect of hydrogen nanobubble addition on combustion characteristics of gasoline engine. *International Journal of Hydrogen Energy* 2013;**38**: 14849–53. doi:10.1016/j.ijhydene.2013.09.063

[11] Şahin Z, Nazım Aksu O, Bayram C. The effects of n-butanol/gasoline blends and 2.5% n-butanol/gasoline blend with 9% water injection into the intake air on the SIE engine performance and exhaust emissions. *Fuel* 2021;**303**. doi:10.1016/j.fuel.2021.121210

[12] Tang Q, Jiang P, Peng C, *et al.* Comparison and analysis of the effects of spark timing and lambda on a high-speed spark ignition engine fuelled with n-butanol/gasoline blends. *Fuel* 2021;**287**: 119505. doi:10.1016/j.fuel.2020.119505

[13] Yu X, Zhao Z, Huang Y, *et al.* Experimental study on the effects of EGR on combustion and emission of an SI engine with gasoline port injection plus ethanol direct injection. *Fuel* 2021;**305**: 121421. doi:10.1016/j.fuel.2021.121421

[14] Zhao L, Qi W, Wang X, *et al.* Potentials of EGR and lean mixture for improving fuel consumption and reducing the emissions of high-proportion butanol-gasoline engines at light load. *Fuel* 2020;**266**: 116959. doi:10.1016/j.fuel.2019.116959

[15] Hodnebrog O, Myhre G, Samset BH, *et al.* Water vapour adjustments and responses differ between climate drivers. *Atmospheric Chemistry and Physics* 2019;**19**: 12887–99. doi:10.5194/acp-19-12887-2019

[16] IPCC. Assessment Report 6 Climate Change 2021: The Physical Science Basis. 2021.

[17] Yang XG, Liu T, Wang CY. Thermally modulated lithium iron phosphate batteries for mass-market electric vehicles. *Nature Energy* 2021;**6**: 176–85. doi:10.1038/s41560-020-00757-7

[18] Brown PT, Caldeira K. Greater future global warming inferred from Earth's recent energy budget. *Nature* 2017;**552**: 45–50. doi:10.1038/nature24672

[19] Davis SJ, Liu Z, Deng Z, *et al.* Carbon dioxide emissions rebound from the COVID-19 pandemic;1–7.

[20] Distribution of greenhouse gas emissions worldwide in 2016, by sub sector. 2021. https://www.statista.com/statistics/1167298/share-ghg-emissions-by-sub-sector-sector-globally/

[21] Abdalla AN, Tao H, Bagaber SA, *et al.* Prediction of emissions and performance of a gasoline engine running with fusel oil–gasoline blends using response surface methodology. *Fuel* 2019;**253**: 1–14. doi:10.1016/j.fuel.2019.04.085

[22] Deng B, Li Q, Chen Y, *et al.* The effect of air/fuel ratio on the CO and NOx emissions for a twin-spark motorcycle gasoline engine under wide range of operating conditions. *Energy* 2019;**169**: 1202–13. doi:10.1016/j.energy.2018.12.113

[23] Sakthivel P, Subramanian KA, Mathai R. Comparative studies on combustion, performance and emission characteristics of a two-wheeler with gasoline and 30% ethanol-gasoline blend using chassis dynamometer. *Applied Thermal Engineering* 2019;**146**: 726–37. doi:10.1016/j.applthermaleng.2018.10.035

[24] Forecast of carbon dioxide emissions worldwide from 2018 to 2050. 2021.https://www.statista.com/statistics/263980/forecast-of-global-carbon-dioxide-emissions/

[25] Hansen AC, Zhang Q, Lyne PWL. Ethanol-diesel fuel blends – A review. *Bioresource Technology* 2005;**96**: 277–85. doi:10.1016/j.biortech.2004.04.007

[26] Akansu SO, Tangöz S, Kahraman N, *et al.* Experimental study of gasoline-ethanol-hydrogen blends combustion in an SI engine. *International Journal of Hydrogen Energy* 2017;**42**: 25781–90. doi:10.1016/j.ijhydene.2017.07.014

[27] Schifter I, Diaz L, Gómez JP, *et al.* Combustion characterization in a single cylinder engine with mid-level hydrated ethanol-gasoline blended fuels. *Fuel* 2013;**103**: 292–8. doi:10.1016/j.fuel.2012.06.002

[28] Wang S, Ji C, Zhang B, *et al.* Performance of a hydroxygen-blended gasoline engine at different hydrogen volume fractions in the hydroxygen. *International Journal of Hydrogen Energy* 2012;**37**: 13209–18. doi:10.1016/j.ijhydene.2012.03.072

[29] White CM, Steeper RR, Lutz AE. The hydrogen-fueled internal combustion engine: a technical review. *International Journal of Hydrogen Energy* 2006;**31**: 1292–305. doi:10.1016/j.ijhydene.2005.12.001

[30] Yilmaz İ, Taştan M. Investigation of hydrogen addition to methanol-gasoline blends in an SI engine. *International Journal of Hydrogen Energy* 2018;**43**: 20252–61. doi:10.1016/j.ijhydene.2018.07.088

[31] Biswal A, Gedam S, Balusamy S, *et al.* Effects of using ternary gasoline-ethanol-LPO blend on PFI engine performance and emissions. *Fuel* 2020;**281**: 118664. doi:10.1016/j.fuel.2020.118664

[32] Taghavifar H, Kaleji BK, Kheyrollahi J. Application of composite TNA nanoparticle with bio-ethanol blend on gasoline fueled SI engine at different lambda ratios. *Fuel* 2020;**277**: 118218. doi:10.1016/j.fuel.2020.118218

[33] Li D, Yu X, Du Y, *et al.* Study on combustion and emissions of a hydrous ethanol/gasoline dual fuel engine with combined injection. *Fuel* 2022;**309**: 122004. doi:10.1016/j.fuel.2021.122004

[34] Sakthivel P, Subramanian KA, Mathai R. Experimental study on unregulated emission characteristics of a two-wheeler with ethanol-gasoline blends (E0 to E50). *Fuel* 2020;**262**: 116504. doi:10.1016/j.fuel.2019.116504

[35] Simsek S, Uslu S. Experimental study of the performance and emissions characteristics of fusel oil/gasoline blends in spark ignited engine using response surface methodology. *Fuel* 2020;**277**: 118182. doi:10.1016/j.fuel.2020.118182

[36] Deng X, Chen Z, Wang X, *et al*. Exhaust noise, performance and emission characteristics of spark ignition engine fuelled with pure gasoline and hydrous ethanol gasoline blends. *Case Studies in Thermal Engineering* 2018;**12**: 55–63. doi:10.1016/j.csite.2018.02.004

[37] Fayyazbakhsh A, Pirouzfar V. Comprehensive overview on diesel additives to reduce emissions, enhance fuel properties and improve engine performance. *Renewable and Sustainable Energy Reviews* 2017;**74**. doi:10.1016/j.rser.2017.03.046

[38] Riaz A, Zahedi G, Klemeš JJ. A review of cleaner production methods for the manufacture of methanol. *Journal of Cleaner Production* 2013;**57**: 19–37. doi:10.1016/j.jclepro.2013.06.017

[39] Al-Hasan M. Effect of ethanol-unleaded gasoline blends on engine performance and exhaust emission. *Energy Conversion and Management* 2003;**44**: 1547–61. doi:10.1016/S0196-8904 (02)00166-8

[40] Najafi G, Ghobadian B, Tavakoli T, *et al*. Performance and exhaust emissions of a gasoline engine with ethanol blended gasoline fuels using artificial neural network. *Applied Energy* 2009;**86**: 630–9. doi:10.1016/j.apenergy.2008.09.017

[41] Yoon SH, Ha SY, Roh HG, *et al*. Effect of bioethanol as an alternative fuel on the emissions reduction characteristics and combustion stability in a spark ignition engine. *Proceedings of the Institution of Mechanical Engineers, Part D: Journal of Automobile Engineering* 2009;**223**: 941–51. doi:10.1243/09544070JAUTO1016

[42] Balat M. Bioethanol as a vehicular fuel: A critical review. *Energy Sources, Part A: Recovery, Utilization and Environmental Effects* 2009;**31**: 1242–55. doi:10.1080/15567030801952334

[43] Li Y, Ning Z, Lee C fon F, *et al*. Effect of acetone-butanol-ethanol (ABE)–gasoline blends on regulated and unregulated emissions in spark-ignition engine. *Energy* 2019;**168**: 1157–67. doi:10.1016/j.energy.2018.12.022

[44] Karavalakis G, Durbin TD, Shrivastava M, *et al*. Impacts of ethanol fuel level on emissions of regulated and unregulated pollutants from a fleet of gasoline light-duty vehicles. *Fuel* 2012;**93**: 549–58. doi:10.1016/j.fuel.2011.09.021

[45] Wu X, Zhang S, Guo X, *et al*. Assessment of ethanol blended fuels for gasoline vehicles in China: Fuel economy, regulated gaseous pollutants and particulate matter. *Environmental Pollution* 2019;**253**: 731–40. doi:10.1016/j.envpol.2019.07.045

[46] Statista. Volume of nitrogen oxides emissions in the United States from 1970 to 2020. 2022. https://www.statista.com/statistics/501284/volume-of-nitrogen-oxides-emissions-us/

[47] Manigandan S, Gunasekar P, Poorchilamban S, *et al*. Effect of addition of hydrogen and TiO2 in gasoline engine in various exhaust gas recirculation ratio. *International Journal of Hydrogen Energy* 2019;**44**: 11205–10. doi:10.1016/j.ijhydene.2019.02.179

[48] Dhyani V, Subramanian KA. Control of backfire and NOx emission reduction in a hydrogen fueled multi-cylinder spark ignition engine using cooled EGR and water injection strategies. *International Journal of Hydrogen Energy* 2019;**44**: 6287–98. doi:10.1016/j. ijhydene.2019.01.129

[49] Agarwal AK, Karare H, Dhar A. Combustion, performance, emissions and particulate characterization of a methanol-gasoline blend (gasohol) fuelled medium duty spark ignition transportation engine. *Fuel Processing Technology* 2014;**121**: 16–24. doi:10.1016/ j.fuproc.2013.12.014

[50] Elfasakhany A. Investigations on the effects of ethanol–methanol–gasoline blends in a spark-ignition engine: Performance and emissions analysis. *Engineering Science and Technology, an International Journal* 2015;**18**: 713–9. doi:10.1016/j.jestch.2015.05.003

[51] Iodice P, Senatore A. Influence of Ethanol-gasoline Blended Fuels on Cold Start Emissions of a Four-stroke Motorcycle. Methodology and Results. 2013. doi:10.4271/2013-24-0117

[52] Wang X, Chen Z, Ni J, *et al.* The effects of hydrous ethanol gasoline on combustion and emission characteristics of a port injection gasoline engine. *Case Studies in Thermal Engineering* 2015;**6**: 147–54. doi:10.1016/j.csite.2015.09.007

[53] Ning Z, Cheung CS, Liu SX. Experimental investigation of the effect of exhaust gas cooling on diesel particulate. *Journal of Aerosol Science* 2004;**35**: 333–45. doi:10.1016/j.jaerosci.2003.10.001

[54] Geng P, Zhang H. Combustion and emission characteristics of a direct-injection gasoline engine using the MMT fuel additive gasoline. *Fuel* 2015;**144**: 380–7. doi:10.1016/j.fuel.2014.12.064

[55] Garrison AW, Lee Wolfe N, Swank RR, *et al.* Environmental fate of methylcyclopentadienyl manganese tricarbonyl. *Environmental Toxicology and Chemistry* 1995;**14**: 1859–64. doi:10.1002/etc.5620141107

[56] Nelson AJ, Reynolds JG, Roos JW. Comprehensive characterization of engine deposits from fuel containing MMT®. *Science of the Total Environment* 2002;**295**: 183–205. doi:10.1016/S0048-9697(02)00093-1

[57] Fayyazbakhsh A, Pirouzfar V. Determining the optimum conditions for modified diesel fuel combustion considering its emission, properties and engine performance. *Energy Conversion and Management* 2016;**113**: 209–19. doi:10.1016/j.enconman.2016.01.058

[58] Valihesari M, Pirouzfar V, Ommi F, *et al.* Investigating the effect of Fe_2O_3 and TiO_2 nanoparticle and engine variables on the gasoline engine performance through statistical analysis. *Fuel* 2019;**254**: 115618. doi:10.1016/j.fuel.2019.115618

[59] Nguyen DD, Moghaddam H, Pirouzfar V, *et al.* Improving the gasoline properties by blending butanol-Al_2O_3 to optimize the engine performance and reduce air pollution. *Energy* 2021;**218**: 119442. doi:10.1016/j.energy.2020.119442

[60] Share of regional deaths attributable to exposure to fine particulate matter (PM2.5) generated by fossil fuel combustion in 2018. 2021.https://www.statista.com/statistics/1203032/fossil-fuel-pollution-deaths-worldwide-by-region/

[61] Wang C, Xu H, Herreros JM, *et al.* Fuel effect on particulate matter composition and soot oxidation in a direct-injection spark ignition (DISI) engine. *Energy and Fuels* 2014;**28**: 2003–12. doi:10.1021/ef402234z

[62] Thakur AK, Kaviti AK, Mehra R, *et al.* Performance analysis of ethanol–gasoline blends on a spark ignition engine: a review. *Biofuels* 2017;**8**: 91–112. doi:10.1080/17597269.2016.1204586

[63] Luo Y, Zhu L, Fang J, *et al.* Size distribution, chemical composition and oxidation reactivity of particulate matter from gasoline direct injection (GDI) engine fueled with ethanol-gasoline fuel. *Applied Thermal Engineering* 2015;**89**: 647–55. doi:10.1016/j.applthermaleng.2015.06.060

[64] Costagliola MA, De Simio L, Iannaccone S, *et al.* Combustion efficiency and engine out emissions of a S.I. engine fueled with alcohol/gasoline blends. *Applied Energy* 2013;**111**: 1162–71. doi:10.1016/j.apenergy.2012.09.042

[65] Li M-D, Wang Z, Zhao Y, *et al.* Experiment Study on Major and Intermediate Species of Ethanol/n-Heptane Premixed Flames. *Combustion Science and Technology* 2013;**185**: 1786–98. doi:10.1080/00102202.2013.839552

[66] Seggiani M, Prati MV, Costagliola MA, *et al.* Bioethanol-gasoline fuel blends: Exhaust emissions and morphological characterization of particulate from a moped engine. *Journal of

the Air and Waste Management Association 2012;**62**: 888–97. doi:10.1080/10962247.2012.671793

[67]   Karavalakis G, Short D, Vu D, et al. The impact of ethanol and iso-butanol blends on gaseous and particulate emissions from two passenger cars equipped with spray-guided and wall-guided direct injection SI (spark ignition) engines. Energy 2015;**82**: 168–79. doi:10.1016/j.energy.2015.01.023

[68]   Ramadhas AS, Singh PK, Sakthivel P, et al. Effect of Ethanol-Gasoline Blends on Combustion and Emissions of a Passenger Car Engine at Part Load Operations. In: SAE Technical Paper. SAE International 2016. doi:10.4271/2016-28-0152

[69]   Matti Maricq M. Soot formation in ethanol/gasoline fuel blend diffusion flames. Combustion and Flame 2012;**159**: 170–80. doi:10.1016/j.combustflame.2011.07.010

[70]   Inal F, Senkan SM. Effects of oxygenate additives on polycyclic aromatic hydrocarbons (PAHs) and soot formation. Combustion Science and Technology 2002;**174**: 1–19. doi:10.1080/00102200290021353

[71]   Li W, Zou C. Deep desulfurization of gasoline by synergistic effect of functionalized B-CD-TiO2-Ag nanoparticles with ionic liquid. Fuel 2018;**227**: 141–9. doi:10.1016/j.fuel.2018.04.083

[72]   Jiang W, Dong L, Liu W, et al. Biodegradable choline-like deep eutectic solvents for extractive desulfurization of fuel. Chemical Engineering and Processing: Process Intensification 2017;**115**: 34–8. doi:10.1016/j.cep.2017.02.004

[73]   Prajapati YN, Verma N. Fixed bed adsorptive desulfurization of thiophene over Cu/Ni-dispersed carbon nanofiber. Fuel 2018;**216**: 381–9. doi:10.1016/j.fuel.2017.11.132

[74]   Menzel R, Iruretagoyena D, Wang Y, et al. Graphene oxide/mixed metal oxide hybrid materials for enhanced adsorption desulfurization of liquid hydrocarbon fuels. Fuel 2016;**181**: 531–6. doi:10.1016/j.fuel.2016.04.125

[75]   Zelekew OA, Kuo DH, Yassin JM, et al. Synthesis of efficient silica supported TiO 2 /Ag 2 O heterostructured catalyst with enhanced photocatalytic performance. Applied Surface Science 2017;**410**: 454–63. doi:10.1016/j.apsusc.2017.03.089

[76]   Lei W, Wenya W, Mominou N, et al. Ultra-deep desulfurization of gasoline through aqueous phase in-situ hydrogenation and photocatalytic oxidation. Applied Catalysis B: Environmental 2016;**193**: 180–8. doi:10.1016/j.apcatb.2016.04.032

[77]   Jahirul MI, Masjuki HH, Saidur R, et al. Comparative engine performance and emission analysis of CNG and gasoline in a retrofitted car engine. Applied Thermal Engineering 2010;**30**: 2219–26. doi:10.1016/j.applthermaleng.2010.05.037

[78]   Shen Y, Li P, Xu X, et al. Selective adsorption for removing sulfur: A potential ultra-deep desulfurization approach of jet fuels. RSC Advances 2012;**2**: 1700–11. doi:10.1039/c1ra00944c

[79]   Yu X, Hao H, Zhang J, et al. Desulfurization of fuel oils by extraction with ionic liquids. Advanced Materials Research 2012;**396–398**: 2221–4. doi:10.4028/www.scientific.net/AMR.396-398.2221

[80]   Si X, Cheng S, Lu Y, et al. Oxidative desulfurization of model oil over Au/Ti-MWW. Catalysis Letters 2008;**122**: 321–4. doi:10.1007/s10562-007-9380-6

[81]   Wang L, Chen Y, Du L, et al. Nickel-heteropolyacids supported on silica gel for ultra-deep desulfurization assisted by Ultrasound and Ultraviolet. Fuel 2013;**105**: 353–7. doi:10.1016/j.fuel.2012.06.021

[82]   Cao R, Zhang X, Wu H, et al. Enhanced pervaporative desulfurization by polydimethylsiloxane membranes embedded with silver/silica core-shell microspheres. Journal of Hazardous Materials 2011;**187**: 324–32. doi:10.1016/j.jhazmat.2011.01.031

[83] Total particulate matter emissions (PM2.5) in the European Union (EU-28) from 2000 to 2019, by sector. 2022.https://www.statista.com/statistics/791153/particulate-matter-emissions-european-union-eu-28/

[84] Chiang TY, Yuan TH, Shie RH, *et al.* Increased incidence of allergic rhinitis, bronchitis and asthma, in children living near a petrochemical complex with SO 2 pollution. *Environ Int* 2016;**96**: 1–7. doi:10.1016/J.ENVINT.2016.08.009

[85] Hoesly RM, Smith SJ, Feng L, *et al.* Historical (1750–2014) anthropogenic emissions of reactive gases and aerosols from the Community Emissions Data System (CEDS). *Geoscientific Model Development* 2018;**11**: 369–408. doi:10.5194/GMD-11-369-2018

[86] Lu Z, Streets DG, Zhang Q, *et al.* Sulfur dioxide emissions in China and sulfur trends in East Asia since 2000. *Atmospheric Chemistry and Physics* 2010;**10**: 6311–31. doi:10.5194/acp-10-6311-2010

[87] Montazeri SM, Sadrnezhaad SK. Kinetics of Sulfur Removal from Tehran Vehicular Gasoline by g-C3N4/SnO2 Nanocomposite. *ACS Omega* 2019;**4**: 13180–8. doi:10.1021/ACSOMEGA.9B01191

[88] Ullah R, Bai P, Wu P, *et al.* Superior performance of freeze-dried Ni/ZnO-Al2O3 adsorbent in the ultra-deep desulfurization of high sulfur model gasoline. *Fuel Processing Technology* 2017;**156**: 505–14. doi:10.1016/J.FUPROC.2016.10.022

[89] Khan NA, Jhung SH. Scandium-Triflate/Metal-Organic Frameworks: Remarkable Adsorbents for Desulfurization and Denitrogenation. *Inorganic Chemistry* 2015;**54**: 11498–504. doi:10.1021/ACS.INORGCHEM.5B02118/SUPPL_FILE/IC5B02118_SI_001.PDF

[90] Bhasarkar JB, Dikshit PK, Moholkar VS. Ultrasound assisted biodesulfurization of liquid fuel using free and immobilized cells of Rhodococcus rhodochrous MTCC 3552: A mechanistic investigation. *Bioresour Technol* 2015;**187**: 369–78. doi:10.1016/J.BIORTECH.2015.03.102

[91] Wang Y, Yang RT. Desulfurization of liquid fuels by adsorption on carbon-based sorbents and ultrasound-assisted sorbent regeneration. *Langmuir* 2007;**23**: 3825–31. doi:10.1021/LA063364Z

[92] Kwon JM, Moon JH, Bae YS, *et al.* Adsorptive desulfurization and denitrogenation of refinery fuels using mesoporous silica adsorbents. *ChemSusChem* 2008;**1**: 307–9. doi:10.1002/CSSC.200700011

[93] Sarda KK, Bhandari A, Pant KK, *et al.* Deep desulfurization of diesel fuel by selective adsorption over Ni/Al2O3 and Ni/ZSM-5 extrudates. *Fuel* 2012;**93**: 86–91. doi:10.1016/J.FUEL.2011.10.020

[94] Gao Y, Han W, Long X, *et al.* Preparation of hydrodesulfurization catalysts using MoS3 nanoparticles as a precursor. *Applied Catalysis B: Environmental* 2018;**224**: 330–40. doi:10.1016/J.APCATB.2017.10.046

[95] Crippa M, Janssens-Maenhout G, Dentener F, *et al.* Forty years of improvements in European air quality: Regional policy-industry interactions with global impacts. *Atmospheric Chemistry and Physics* 2016;**16**: 3825–41. doi:10.5194/ACP-16-3825-2016

[96] Gouw JA de, Parrish DD, Frost GJ, *et al.* Reduced emissions of CO2, NOx, and SO2 from U.S. power plants owing to switch from coal to natural gas with combined cycle technology. *Earth's Future* 2014;**2**: 75–82. doi:10.1002/2013EF000196

[97] Zheng B, Tong D, Li M, *et al.* Trends in China's anthropogenic emissions since 2010 as the consequence of clean air actions. *Atmospheric Chemistry and Physics* 2018;**18**: 14095–111. doi:10.5194/ACP-18-14095-2018

[98] Zhong Q, Shen H, Yun X, *et al.* Global Sulfur Dioxide Emissions and the Driving Forces. *Environmental Science and Technology* 2020;**54**: 6508–17. doi:10.1021/ACS.EST.9B07696

[99] Badra J, AlRamadan AS, Sarathy SM. Optimization of the octane response of gasoline/ethanol blends. *Applied Energy* 2017;**203**: 778–93. doi:10.1016/j.apenergy.2017.06.084

[100] Jiang C, Huang G, Liu G, *et al.* Optimizing gasoline compression ignition engine performance and emissions: Combined effects of exhaust gas recirculation and fuel octane number. *Applied Thermal Engineering* 2019;**153**: 669–77. doi:10.1016/j.applthermaleng.2019.03.054

[101] Wang C, Li Y, Xu C, *et al.* Methanol as an octane booster for gasoline fuels. *Fuel* 2019;**248**: 76–84. doi:10.1016/j.fuel.2019.02.128

[102] Morganti K, Viollet Y, Head R, *et al.* Maximizing the benefits of high octane fuels in spark-ignition engines. *Fuel* 2017;**207**: 470–87. doi:10.1016/j.fuel.2017.06.066

[103] Petre MN, Rosca P, Dragomir RE, *et al.* Bioalcohols – Compounds for reformulated gasolines II. Prediction of volatility properties for fuel-alcohols blends. *Revista de Chimie* 2010;**61**: 805–8.

[104] Christensen E, Yanowitz J, Ratcliff M, *et al.* Renewable oxygenate blending effects on gasoline properties. *Energy and Fuels* 2011;**25**: 4723–33. doi:10.1021/ef2010089

[105] Dhamodaran G, Esakkimuthu GS. Experimental measurement of physico-chemical properties of oxygenate (DIPE) blended gasoline. *Measurement: Journal of the International Measurement Confederation* 2019;**134**: 280–5. doi:10.1016/j.measurement.2018.10.077

[106] Rodríguez-Antón LM, Gutiérrez-Martín F, Hernández-Campos M. Physical properties of gasoline-ETBE-isobutanol (in comparison with ethanol) ternary blends and their impact on regulatory compliance. *Energy* 2019;**185**: 68–76. doi:10.1016/j.energy.2019.07.050

[107] Rodríguez-Antón LM, Gutiérrez-Martín F, Martinez-Arevalo C. Experimental determination of some physical properties of gasoline, ethanol and ETBE ternary blends. *Fuel* 2015;**156**: 81–6. doi:10.1016/j.fuel.2015.04.040

5 The influence of other types of additives blended with gasoline

5.1 Brake-specific fuel consumption

The heating value of alcohol, especially methanol and ethanol, is lower than that of gasoline. Henceforth, more fuel is needed to get the same engine power output. This feature shows that the ethanol's heating value blended with gasoline fuel reduces by enhancing alcohol content. According to the results, cerium oxide and aluminum nanoparticles, due to faster evaporation rate [1, 2], cause decreased BSFC when added to the fuel because of complete combustion and emission of the nitrogen produced from the reaction of nanoparticles in the ignition process. This result was achieved when the amount of ethanol heat had been 1.3 times less than the base fuel [3]. Jahirul and his co-workers [4] worked on the influence of the blending of Al_2O_3/TiO_2 nano-materials with gasoline on engine performance at different speeds and throttle valve positions. They found that at high engine speed, the BSFC increased, especially at low throttle. Thus, there is no considerable fluctuation in the BSFC from 50% to 100% of throttle valve openings.

In this context, additional research has been done in which it was discovered that nanoparticles positively reduce BSFC when they are used in certain concentrations and amounts. This finding is relevant because it was found that nanoparticles have a positive effect on lowering BSFC. As an illustration, in order to name a few of the most effective nanoparticles: TiO_2 (−23.42%), Mn_2O_3 (−38.89%), CeO_2 (−30%), and graphene oxide (GO) (−20%) [5].

Recent research has revealed that lowering fuel consumption may be accomplished by utilizing additives such as graphite nanoparticles, Fe_2O_3 nanoparticles, and tire pyrolysis oil (TPO) at varying proportions. These additives have proven to be quite efficient in this endeavor. Graphite nanoparticles at concentrations of 40, 80, and 120 mg/L, as well as Fe_2O_3 nanoparticles and both 5% and 10% TPO, were combined with gasoline in one experiment and put through their paces under a set of predetermined operating parameters on a SI engine. According to the findings, utilizing nanoparticles and combining them with fossil fuels not only improves the efficacy of relevant indicators but also lowers BSFC. This was demonstrated by the study's findings [6].

Oh et al. [7] studied the influence of adding hydrogen nano-bubble (HNB) to fuel on engine performance improvement. They concluded that the fuel consumption rate showed a smooth enhancement with increasing engine load.

https://doi.org/10.1515/9783110999969-005

5.2 Brake thermal efficiency and power

BTE is a critical factor in choosing an oxygenated or nanoparticle additive. Blending additives cause an increase in the BTE due to the higher latent vaporization heat. He et al. [8] investigated the effect of adding normal butanol to gasoline on combustion characteristics in the in-cylinder direct injection (DI) engine and the port injection (PI). They concluded that the combustion time was shortened, engine ignition time was advanced, and BTE was improved only by adding n-butanol. Yusoff et al. [9] suggested that the brake power is improved in an almost linear fashion when the engine speed increases. Moreover, they concluded that most blends have no considerable result of brake power. The normal-Bu20 and iso-Bu20 blend have higher brake power, whereas gasoline reflects a negligible increment of 1.36% and 1.73%, respectively. Since iso-butanol and normal-butanol are essentially oxygenated when blended, the octane rating is increased. They may not be prone to auto-ignition, but they can improve the brake power of blended fuel. BTE is increased by fueling hydrous ethanol-gasoline instead of pure gasoline at the same tested engine speed. The main reason for such influence is associated with the fact that ethanol produces the hydroxyl radical (−OH) and leads the combustion to be more complete and improves flame propagation speed. Therefore, lower heat loss to the cylinder walls and shorter combustion periods are obtained, and BTE is enhanced. Sebayang et al. [10] suggested that the BTE of bioethanol-gasoline is higher than neat gasoline due to the high mean effective pressure of the blended fuel. Oh et al. [7] studied the influence of adding HNB additive to gasoline on engine performance. Their results indicated a direct dependency between BTE and engine load because of the increase in the air-fuel rate. Also, they reported that BTE in the HNB gasoline combustion is higher than neat gasoline fuel at all engine loads.

When GO nanoparticles, which may boost BTE by up to 17%, as well as graphene nanoplatelet (GNP) nanoparticles were added to the fuel and examined, it was shown that GNP greatly minimizes ignition delay ID, which is one of the most critical elements in determining the quality of combustion. On the other hand, the higher thermal conductivity of GNP contributes to an improvement in BTE [11]. GO has a lower thermal conductivity than GNP, which causes quicker burning. Still, GNP enhances BTE owing to its high thermal conductivity is one of the noteworthy outcomes that have been gained from this experiment [11]. The findings of other tests conducted by different researchers on nanoparticle additions and their influence on BTE hint at the notion that a factor known as a calorific value directly affects BTE. Because the rapid evaporation of nanoparticles owing to complete combustion helps to transmit heat between fuel components and nanoparticles more quickly, the BTE can be improved by mixing fuels with greater calorific values and higher viscosities [2, 12, 13]. On the other hand, the ratio between the surface area of nanoparticle droplets and their total volume is directly related to the rate of evaporation, which, as was previously mentioned, has a positive effect on reducing ignition delay and combustion completion, leading to an increase in BTE for nanoparticles composed of TiO_2/SiO_2, CeO_2, and GO [14–16].

Extensive research has been done in the field of nanoparticles and diesel. According to the findings of this research, cerium oxide nanoparticles are a suitable additive for fuel and increase thermal efficiency due to their powerful catalytic properties. These properties result from their high surface-to-volume ratio and their high thermal stability. Because the nanoparticles are transported quicker with rising heat, which enhances the pace of combustion reactions, the presence of cerium oxide in the fuel decreases the ignition delay and evaporation time. This is because the rate of combustion processes rises. Therefore, the presence of oxygen in this nanoparticle serves as an accelerator, which results in high BTE efficiency [18].

5.3 NOx emissions

Previous research [19, 20] reported that newer engines could better control NO_x emissions when using ethanol down to its better-sophisticated engine control strategies and after-treatment devices. Manigandan et al. [21] concluded that different ratios of hydrogen and exhaust gas recirculation (EGR) and TiO_2 5% decrease NOX, CO, and CO_2 emissions due to the effect of hydrogenated fuel. Hydrogen reduces flame development, rapid combustion, and combustion duration while using EGR increases the above parameters. Vipin and Subramanian [22] reported that water injection efficiently affects emission and knocking compared to EGR.

Moreover, other research demonstrated that NOx emissions and the knocking effect are reduced by adding EGR [23, 24]. Besides, nanoparticles can separate the oxygen existing in the water. However, the presence of oxygen makes combustion complete [25–27]. After adding water and TiO_2 nanoparticles, it is concluded that NOx is decreased, but combustion tends to be complete.

It has been demonstrated that adding ethanol to gasoline results in an increase in NOx emissions as a consequence of an increase in oxygen concentration as well as oxygen bonding and the oxidation process. On the other hand, when we use nanoparticles with fuel, we experience a reduction in NOx emissions. For instance, as shown in Figure 5.1, when we use a mixture of gasoline that contains 10% ethanol and 10 parts per million Mn_2O_3, we experience a reduction in NOx emissions of 23.43%. Additionally, when we increase the amount of Mn_2O_3 nanoparticles that we use as fuel and replace compound gasoline that contains gasoline-10% ethanol-20 ppm Mn_2O_3, we experience a 32.34% reduction of NOx [28].

As was noted, the presence of ethanol causes an increase in NOx emissions; nevertheless, a recent study shows that bio-ethanol can lower NOx while simultaneously raising octane number. Because of this, we will see a drop in flame temperature due to the delay in the combustion process. However, some nanoparticles are known to enhance NOx emissions. One example of this is the TNA nanocomposite. If a fuel that

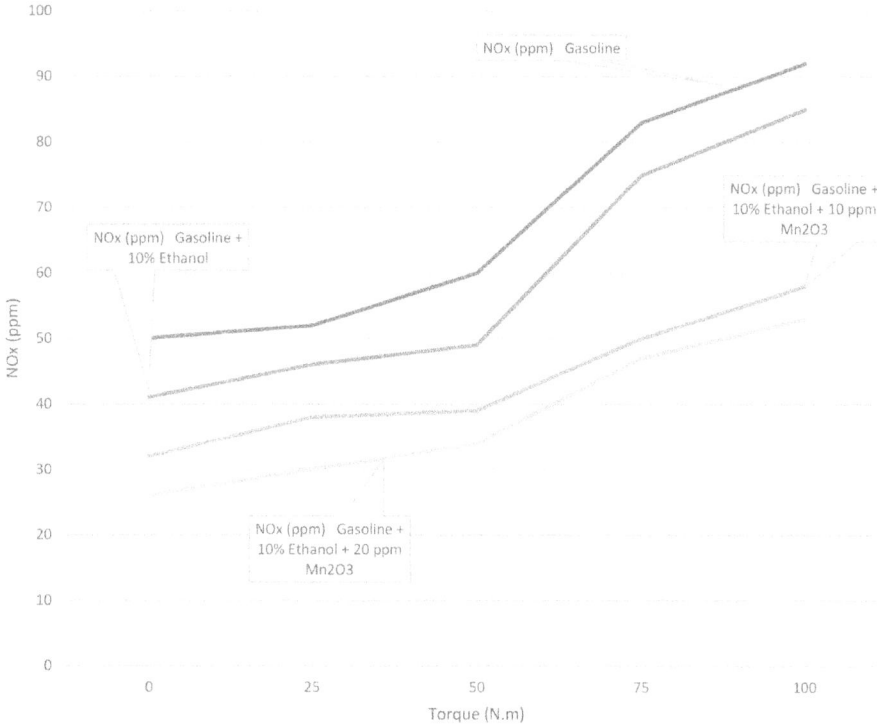

Figure 5.1: The effects of gasoline, ethanol, and manganese oxdide on NOx emissions.

contains bioethanol is utilized, there is a possibility that the concentration of NOx will slightly rise during the combustion process and then persist in the exhaust. As a result of using the TNA nanocomposite, this material can break the bonds in the NOx and release oxygen, which ultimately increases the NOx emissions [29–31].

5.4 CO emissions

As CO emissions are one of the most dangerous pollutants, researchers pay more attention to it to make the highest possible reduction compared to some pollutants such as HC, CO_2, and NOx. In this regard, more techniques have been used to reduce this emission from burning gasoline. One of them uses a newer catalyst system, such as three-way catalysts, whose performance is improving continuously and is used to oxidize CO to CO_2 [32–34]. The rise in oxygen content that occurs with the combustion of oxygenated fuels is the root cause of CO emissions. As a result, one strategy for reducing carbon monoxide emissions is to cut back on the quantity of oxygen included in fuels and additives. Because of this, the adoption of a CNT multilayer pipe construction is one of these ways. This technology has the potential to be successful in absorbing oxygen and lowering CO emissions [35, 36]. Since the incomplete combustion of fossil fuel causes CO emissions, and since this occurs most frequently with diesel fuel, a significant amount of study has been done on this subject to find a solution to this issue. Research has demonstrated that incorporating nano-catalysts such as alumina into biodiesel fuel B5 (95% diesel + 5% biodiesel) and B10 (90% diesel + 10% biodiesel) can be beneficial in terms of lowering CO emissions by 2.94%. Since it has such a large contact surface, alumina boosts the amount of chemical reactivity, which in turn shortens the time it takes for combustion to begin. In the same way, the degree to which fuel-air mix improves ultimately results in more thorough and efficient combustion. On the other hand, lowering the viscosity of the fuel plays a significant part in increasing its atomization, and all of these processes ultimately result in lower CO emissions [37, 38].

5.5 HC emissions

Additionally, the HC emissions bring about other products like particulate nucleation. It occurs at high exhaust gas temperatures [39]. Geng et al. [40] added methylcyclopentadienyl manganese tricarbonyl (MMT) to gasoline to detect the fuel emission. This study used a gasoline direct injection engine with a speed of 2,000 rpm and the EURO V ultralow sulfur gasoline. Compared with other additives, the steam pressure of MMT is lower, and it is also less soluble [41]. Based on the results and tables, it is inferred that the HC emissions are remarkably diminished because it is homogenously blended, and the combustion efficiency is increased [42]. As mentioned earlier, nanoparticles mainly reduce fuel emissions [43]. They may also cause converse outcomes. For instance, Valihesari et al. [44] analyzed the Fe_2O_3 and TiO_2 nanoparticles as a gasoline additive in their research. They indicated that the flame silencing delay is increased while reaching the cylinder wall at the engine's highest speed for the high temperature of the combustion chamber. The TiO_2 nano-particles additive decreases the HC emissions, whereas the Fe_2O_3 nano-particles enhance the HC emission because incomplete combustion causes emissions at higher speeds.

5.6 Soot emissions

Li and Zou [45] focused on Ag and β-cyclodextrins, which were related to 5% of the nanoparticles of TiO_2 to produce CD-TiO_2-Agβ. On the other hand, limiting the quantity of sulfur and gasoline emissions is counted as an essential priority in the petroleum refining industry. Different methods sulfurize gasoline. Non-hydrodesulfurization is regarded as a fast and economical technology. This method contains catalytic hydrodesulfurization, fluid catalytic cracking (FCC), photocatalytic oxidation, oxidative biological, and adsorptive desulfurization [46, 47]. TiO_2 comprehensively wipes out environmental pollution. It is also noteworthy because it is cheap, nontoxic, and durable.

Furthermore, it has unique optical specifications, catalytic and photoelectric conversion potential, and properties. It is also a metallic oxide that acts as an absorbent in sulfurization processes [48, 49]. Wang et al. [50] deliberated the blending of ZSM-5 with ZrO_2, titanium, and NiO for sulfurization of FCC.

Considering the limits of HDS, other technologies such as bio-desulfurization [3, 4], selective adsorption [51], physical extraction [52], oxidative desulfurization (ODS) [53, 54] complexation, and pervaporation [55] can be applied to eliminate the thiophene compounds. Among the items mentioned above, it is turned out that ODS is the most suitable way for gasoline sulfurization in FCC. It is a good option for a high percentage of sulfurization and not losing the octane number.

Catalyst is vital in the oxidation of soot and CO emissions [56, 57]. Regarding soot oxidation, transition metal oxides such as CeO_2, Fe_2O_3, Co_3O_4, and MnOx [58–60], rare earth metals oxides [61, 62], and precious metals (that are the most active catalysts but at the highest price) [63, 64] have been used in a variety range.

5.7 Octane number

The influence of nano additives on octane number could be different due to the used additive. Saxena et al. [65] added methanol to gasoline with improved catalyst life on nano-crystalline ZSM-5 catalysts. They reported that this blend subjects to develop the octane number and reduce undesired benzene. Viswanadham et al. [66] investigated the effect of using a zeolite-based catalyst for octane number enhancement. They suggested that this catalyst increased the fuel octane number. They also clarified considerations on the size distribution of pores, metal, and acid sites in the catalyst. They claimed that these are suitable for the significant long-term performance regarding boosting the octane. Also, this catalyst has the potential to improve other fuel properties.

Research conducted by a number of scientists demonstrates that increasing the proportion of multi-walled amido-functionalized carbon nanotubes (MWNT) in gasoline results in a higher-octane rating for the fuel. In addition, the impact that octadecylamine has when utilized for amido-functionalized functionalization of MWNT is more significant for boosting the octane number of gasoline. In addition to dodecylamine, increasing the number of additives causes the octane number to rise. To counter this, add on should be up to 7 ppm among the additives since, above this level, the solubility of functionalized MWNTs is not permanent. Because of this, the maximum quantity that it should be is 7 ppm [67]. Carbon nanotubes (CNTs), capable of scavenging free radicals, are appropriate anti-knock additives. This feature makes them useful in addition to improving fuel atomization. CNTs can have more tremendous potential for reactivity if they are functionalized with amide groups, which are compounds that can be dissolved in gasoline to raise that fuel's octane rating [68].

References

[1]   Soukht Saraee H, Jafarmadar S, Taghavifar H, *et al*. Reduction of emissions and fuel consumption in a compression ignition engine using nanoparticles. *International Journal of Environmental Science and Technology* 2015;**12**: 2245–52. doi:10.1007/s13762-015-0759-4

[2]   Srinivasa Rao M, Anand RB. Performance and emission characteristics improvement studies on a biodiesel fuelled DICI engine using water and AlO(OH) nanoparticles. *Applied Thermal Engineering* 2016;**98**: 636–45. doi:10.1016/J.APPLTHERMALENG.2015.12.090

[3]   Iodice P, Senatore A. Influence of Ethanol-gasoline Blended Fuels on Cold Start Emissions of a Four-stroke Motorcycle. Methodology and Results. 2013. doi:10.4271/2013-24-0117

[4]   Jahirul MI, Masjuki HH, Saidur R, *et al*. Comparative engine performance and emission analysis of CNG and gasoline in a retrofitted car engine. *Applied Thermal Engineering* 2010;**30**: 2219–26. doi:10.1016/j.applthermaleng.2010.05.037

[5]   Hatami M, Hasanpour M, Jing D. Recent developments of nanoparticles additives to the consumables liquids in internal combustion engines: Part I: Nano-fuels. *Journal of Molecular Liquids* 2020;**318**: 114250. doi:10.1016/J.MOLLIQ.2020.114250

[6]   Yaqoob H, Teoh YH, Sher F, *et al*. Energy, exergy, sustainability and economic analysis of waste tire pyrolysis oil blends with different nanoparticle additives in spark ignition engine. *Energy* 2022;**251**: 123697. doi:10.1016/J.ENERGY.2022.123697

[7]   Oh SH, Yoon SH, Song H, *et al*. Effect of hydrogen nanobubble addition on combustion characteristics of gasoline engine. *International Journal of Hydrogen Energy* 2013;**38**: 14849–53. doi:10.1016/j.ijhydene.2013.09.063

[8]   He BQ, Chen X, Lin CL, *et al*. Combustion characteristics of a gasoline engine with independent intake port injection and direct injection systems for n-butanol and gasoline. *Energy Conversion and Management* 2016;**124**: 556–65. doi:10.1016/j.enconman.2016.07.053

[9]   Yusoff MNAM, Zulkifli NWM, Masjuki HH, *et al*. Performance and emission characteristics of a spark ignition engine fuelled with butanol isomer-gasoline blends. *Transportation Research Part D: Transport and Environment* 2017;**57**: 23–38. doi:10.1016/j.trd.2017.09.004

[10]   Sebayang AH, Masjuki HH, Ong HC, *et al*. Prediction of engine performance and emissions with Manihot glaziovii bioethanol – Gasoline blended using extreme learning machine. *Fuel* 2017;**210**: 914–21. doi:10.1016/j.fuel.2017.08.102

[11]   Chacko N, Jeyaseelan T. Comparative evaluation of graphene oxide and graphene nanoplatelets as fuel additives on the combustion and emission characteristics of a diesel engine fuelled with diesel and biodiesel blend. *Fuel Processing Technology* 2020;**204**: 106406. doi:10.1016/J.FUPROC.2020.106406

[12]   Dhinesh B, Maria Ambrose Raj, Y, Kalaiselvan C, *et al*. A numerical and experimental assessment of a coated diesel engine powered by high-performance nano biofuel. *Energy Conversion and Management* 2018;**171**: 815–24. doi:10.1016/J.ENCONMAN.2018.06.039

[13]   Kumaran P, Joel Godwin A, Amirthaganesan S. Effect of microwave synthesized hydroxyapatite nanorods using Hibiscus rosa-sinensis added waste cooking oil (WCO) methyl ester biodiesel blends on the performance characteristics and emission of a diesel engine. *Materials Today: Proceedings* 2020;**22**: 1047–53. doi:10.1016/J.MATPR.2019.11.288

[14]   Karthikeyan S, Prathima A. Environmental effect of CI engine using microalgae methyl ester with doped nano additives. *Transportation Research Part D: Transport and Environment* 2017;**50**: 385–96. doi:10.1016/J.TRD.2016.11.028

[15]   Kumaravel ST, Murugesan A, Vijayakumar C, *et al.* Enhancing the fuel properties of tyre oil diesel blends by doping nano additives for green environments. *Journal of Cleaner Production* 2019;**240**: 118128. doi:10.1016/J.JCLEPRO.2019.118128

[16]   Gowtham M, Prakash R. Control of regulated and unregulated emissions from a CI engine using reformulated nano fuel emulsions. *Fuel* 2020;**271**: 117596. doi:10.1016/J. FUEL.2020.117596

[17]   Zamankhan F, Pirouzfar V, Ommi F, *et al.* Investigating the effect of MgO and CeO2 metal nanoparticle on the gasoline fuel properties: empirical modeling and process optimization by surface methodology. *Environmental Science and Pollution Research* 2018;**25**: 22889–902. doi:10.1007/s11356-018-2066-3

[18]   Nanthagopal K, Kishna RS, Atabani AE, *et al.* A compressive review on the effects of alcohols and nanoparticles as an oxygenated enhancer in compression ignition engine. *Energy Conversion and Management* 2020;**203**: 112244. doi:10.1016/J.ENCONMAN.2019.112244

[19]   Karavalakis G, Durbin TD, Shrivastava M, *et al.* Impacts of ethanol fuel level on emissions of regulated and unregulated pollutants from a fleet of gasoline light-duty vehicles. *Fuel* 2012;**93**: 549–58. doi:10.1016/j.fuel.2011.09.021

[20]   Wu X, Zhang S, Guo X, *et al.* Assessment of ethanol blended fuels for gasoline vehicles in China: Fuel economy, regulated gaseous pollutants and particulate matter. *Environmental Pollution* 2019;**253**: 731–40. doi:10.1016/j.envpol.2019.07.045

[21]   Manigandan S, Gunasekar P, Poorchilamban S, *et al.* Effect of addition of hydrogen and TiO2 in gasoline engine in various exhaust gas recirculation ratio. *International Journal of Hydrogen Energy* 2019;**44**: 11205–18. doi:10.1016/j.ijhydene.2019.02.179

[22]   Dhyani V, Subramanian KA. Control of backfire and NOx emission reduction in a hydrogen fueled multi-cylinder spark ignition engine using cooled EGR and water injection strategies. *International Journal of Hydrogen Energy* 2019;**44**: 6287–98. doi:10.1016/j. ijhydene.2019.01.129

[23]   Dimitriou P, Kumar M, Tsujimura T, *et al.* Combustion and emission characteristics of a hydrogen-diesel dual-fuel engine. *International Journal of Hydrogen Energy* 2018;**43**: 13605–17. doi:10.1016/j.ijhydene.2018.05.062

[24]   Luo Q he, Hu J Bin, Sun B gang, *et al.* Experimental investigation of combustion characteristics and NOx emission of a turbocharged hydrogen internal combustion engine. *International Journal of Hydrogen Energy* 2019;**44**: 5573–84. doi:10.1016/j. ijhydene.2018.08.184

[25]   Fayyazbakhsh A, Pirouzfar V. Comprehensive overview on diesel additives to reduce emissions, enhance fuel properties and improve engine performance. *Renewable and Sustainable Energy Reviews* 2017;**74**. doi:10.1016/j.rser.2017.03.046

[26]   Işik MZ, Aydin H. Investigation on the effects of gasoline reactivity controlled compression ignition application in a diesel generator in high loads using safflower biodiesel blends. *Renewable Energy* 2019;**133**: 177–89. doi:10.1016/j.renene.2018.10.025

[27]   Ratcliff MA, Windom B, Fioroni GM, *et al.* Impact of ethanol blending into gasoline on aromatic compound evaporation and particle emissions from a gasoline direct injection engine. *Applied Energy* 2019;**250**: 1618–31. doi:10.1016/j.apenergy.2019.05.030

[28]   Amirabedi M, Jafarmadar S, Khalilarya S. Experimental investigation the effect of Mn2O3 nanoparticle on the performance and emission of SI gasoline fueled with mixture of ethanol and gasoline. *Applied Thermal Engineering* 2019;**149**: 512–9. doi:10.1016/J. APPLTHERMALENG.2018.12.058

[29]   Taghavifar H, Kaleji BK, Kheyrollahi J. Application of composite TNA nanoparticle with bio-ethanol blend on gasoline fueled SI engine at different lambda ratios. *Fuel* 2020;**277**: 118218. doi:10.1016/J.FUEL.2020.118218

[30]   Chen AF, Akmal Adzmi M, Adam A, *et al.* Combustion characteristics, engine performances and emissions of a diesel engine using nanoparticle-diesel fuel blends with aluminium oxide, carbon nanotubes and silicon oxide. *Energy Conversion and Management* 2018;**171**: 461–77. doi:10.1016/J.ENCONMAN.2018.06.004

[31]   Ma Y, Zhu M, Zhang D. The effect of a homogeneous combustion catalyst on exhaust emissions from a single cylinder diesel engine. *Applied Energy* 2013;**102**: 556–62. doi:10.1016/J.APENERGY.2012.08.028

[32]   Liu H, Li Z, Zhang M, *et al.* Exhaust non-volatile particle filtration characteristics of three-way catalyst and influencing factors in a gasoline direct injection engine compared to gasoline particulate filter. *Fuel* 2021;**290**: 120065. doi:10.1016/j.fuel.2020.120065

[33]   Velmurugan D V, Mckelvey T, Olsson J. simulation framework for simulation framework for simulation framework for A simulation framework evaluation of a gasoline engine equipped evaluation of a gasoline engine equipped A simulation framework for cold-start evaluation of a gasoline engine equip. *IFAC PapersOnLine* 2021;**54**: 526–33. doi:10.1016/j.ifacol.2021.10.216

[34]   Herreros JM, Oliva F, Zeraati-rezaei S, *et al.* Effects of high octane additivated gasoline fuel on Three Way Catalysts performance under an accelerated catalyst ageing procedure CO EU MON. *Fuel* 2022;**312**: 122970. doi:10.1016/j.fuel.2021.122970

[35]   Mei D, Zuo L, Adu-Mensah D, *et al.* Combustion characteristics and emissions of a common rail diesel engine using nanoparticle-diesel blends with carbon nanotube and molybdenum trioxide. *Applied Thermal Engineering* 2019;**162**: 114238. doi:10.1016/J.APPLTHERMALENG.2019.114238

[36]   Khond VW, Kriplani VM. Effect of nanofluid additives on performances and emissions of emulsified diesel and biodiesel fueled stationary CI engine: A comprehensive review. *Renewable and Sustainable Energy Reviews* 2016;**59**: 1338–48. doi:10.1016/J.RSER.2016.01.051

[37]   Safieddin Ardebili SM, Taghipoor A, Solmaz H, *et al.* The effect of nano-biochar on the performance and emissions of a diesel engine fueled with fusel oil-diesel fuel. *Fuel* 2020;**268**: 117356. doi:10.1016/J.FUEL.2020.117356

[38]   Hosseini SH, Taghizadeh-Alisaraei A, Ghobadian B, *et al.* Effect of added alumina as nano-catalyst to diesel-biodiesel blends on performance and emission characteristics of CI engine. *Energy* 2017;**124**: 543–52. doi:10.1016/J.ENERGY.2017.02.109

[39]   Ning Z, Cheung CS, Liu SX. Experimental investigation of the effect of exhaust gas cooling on diesel particulate. *Journal of Aerosol Science* 2004;**35**: 333–45. doi:10.1016/j.jaerosci.2003.10.001

[40]   Geng P, Zhang H. Combustion and emission characteristics of a direct-injection gasoline engine using the MMT fuel additive gasoline. *Fuel* 2015;**144**: 380–7. doi:10.1016/j.fuel.2014.12.064

[41]   Garrison AW, Lee Wolfe N, Swank RR, *et al.* Environmental fate of methylcyclopentadienyl manganese tricarbonyl. *Environmental Toxicology and Chemistry* 1995;**14**: 1859–64. doi:10.1002/etc.5620141107

[42]   Nelson AJ, Reynolds JG, Roos JW. Comprehensive characterization of engine deposits from fuel containing MMT®. *Science of the Total Environment* 2002;**295**: 183–205. doi:10.1016/S0048-9697(02)00093-1

[43] Fayyazbakhsh A, Pirouzfar V. Determining the optimum conditions for modified diesel fuel combustion considering its emission, properties and engine performance. *Energy Conversion and Management* 2016;**113**: 209–19. doi:10.1016/j.enconman.2016.01.058

[44] Valihesari M, Pirouzfar V, Ommi F, *et al*. Investigating the effect of Fe2O3 and TiO2 nanoparticle and engine variables on the gasoline engine performance through statistical analysis. *Fuel* 2019;**254**: 115618. doi:10.1016/j.fuel.2019.115618

[45] Li W, Zou C. Deep desulfurization of gasoline by synergistic effect of functionalized B-CD-TiO2 -Ag nanoparticles with ionic liquid. *Fuel* 2018;**227**: 141–9. doi:10.1016/j.fuel.2018.04.083

[46] Jiang W, Dong L, Liu W, *et al*. Biodegradable choline-like deep eutectic solvents for extractive desulfurization of fuel. *Chemical Engineering and Processing: Process Intensification* 2017;**115**: 34–8. doi:10.1016/j.cep.2017.02.004

[47] Prajapati YN, Verma N. Fixed bed adsorptive desulfurization of thiophene over Cu/Ni-dispersed carbon nanofiber. *Fuel* 2018;**216**: 381–9. doi:10.1016/j.fuel.2017.11.132

[48] Menzel R, Iruretagoyena D, Wang Y, *et al*. Graphene oxide/mixed metal oxide hybrid materials for enhanced adsorption desulfurization of liquid hydrocarbon fuels. *Fuel* 2016;**181**: 531–6. doi:10.1016/j.fuel.2016.04.125

[49] Zelekew OA, Kuo DH, Yassin JM, *et al*. Synthesis of efficient silica supported TiO 2 /Ag 2 O heterostructured catalyst with enhanced photocatalytic performance. *Applied Surface Science* 2017;**410**: 454–63. doi:10.1016/j.apsusc.2017.03.089

[50] Lei W, Wenya W, Mominou N, *et al*. Ultra-deep desulfurization of gasoline through aqueous phase in-situ hydrogenation and photocatalytic oxidation. *Applied Catalysis B: Environmental* 2016;**193**: 180–8. doi:10.1016/j.apcatb.2016.04.032

[51] Shen Y, Li P, Xu X, *et al*. Selective adsorption for removing sulfur: A potential ultra-deep desulfurization approach of jet fuels. *RSC Advances* 2012;**2**: 1700–11. doi:10.1039/c1ra00944c

[52] Yu X, Hao H, Zhang J, *et al*. Desulfurization of fuel oils by extraction with ionic liquids. *Advanced Materials Research* 2012;**396–398**: 2221–4. doi:10.4028/www.scientific.net/AMR.396-398.2221

[53] Si X, Cheng S, Lu Y, *et al*. Oxidative desulfurization of model oil over Au/Ti-MWW. *Catalysis Letters* 2008;**122**: 321–4. doi:10.1007/s10562-007-9380-6

[54] Wang L, Chen Y, Du L, *et al*. Nickel-heteropolyacids supported on silica gel for ultra-deep desulfurization assisted by Ultrasound and Ultraviolet. *Fuel* 2013;**105**: 353–7. doi:10.1016/j.fuel.2012.06.021

[55] Cao R, Zhang X, Wu H, *et al*. Enhanced pervaporative desulfurization by polydimethylsiloxane membranes embedded with silver/silica core-shell microspheres. *Journal of Hazardous Materials* 2011;**187**: 324–32. doi:10.1016/j.jhazmat.2011.01.031

[56] Yao P, He J, Jiang X, *et al*. Factors determining gasoline soot abatement over CeO2–ZrO2-MnOx catalysts under low oxygen concentration condition. *Journal of the Energy Institute* 2020;**93**: 774–83. doi:10.1016/j.joei.2019.05.005

[57] Hernández WY, Tsampas MN, Zhao C, *et al*. La/Sr-based perovskites as soot oxidation catalysts for Gasoline Particulate Filters. *Catalysis Today* 2015;**258**: 525–34. doi:10.1016/j.cattod.2014.12.021

[58] Lee JH, Lee SH, Choung JW, *et al*. Ag-incorporated macroporous CeO2 catalysts for soot oxidation: Effects of Ag amount on the generation of active oxygen species. *Applied Catalysis B: Environmental* 2019;**246**: 356–66. doi:10.1016/j.apcatb.2019.01.064

[59] Jin B, Zhao B, Liu S, *et al*. SmMn2O5 catalysts modified with silver for soot oxidation: Dispersion of silver and distortion of mullite. *Applied Catalysis B: Environmental* 2020;**273**. doi:10.1016/j.apcatb.2020.119058

[60] Lee JH, Lee BJ, Lee DW, *et al*. Synergistic effect of Cu on a Ag-loaded CeO2 catalyst for soot oxidation with improved generation of active oxygen species and reducibility. *Fuel* 2020;**275**: 117930. doi:10.1016/j.fuel.2020.117930

[61] Ali S, Wu X, Zuhra Z, *et al*. Cu-Mn-Ce mixed oxides catalysts for soot oxidation and their mechanistic chemistry. *Applied Surface Science* 2020;**512**: 145602. doi:10.1016/j. apsusc.2020.145602

[62] Rao C, Liu R, Feng X, *et al*. Three-dimensionally ordered macroporous SnO2-based solid solution catalysts for effective soot oxidation. *Cuihua Xuebao/Chinese Journal of Catalysis* 2018;**39**: 1683–94. doi:10.1016/S1872-2067(18)63123-7

[63] Lee S, Lee H, Song C, *et al*. Experimental study on fundamental effect of H2 for catalytic soot oxidation with Pt/CeO2 using a flow reactor system. *Journal of the Energy Institute* 2019;**92**: 1419–27. doi:10.1016/j.joei.2018.08.009

[64] Gao Y, Duan A, Liu S, *et al*. Study of Ag/CexNd1-xO2 nanocubes as soot oxidation catalysts for gasoline particulate filters: Balancing catalyst activity and stability by Nd doping. *Applied Catalysis B: Environmental* 2017;**203**: 116–26. doi:10.1016/j.apcatb.2016.10.006

[65] Saxena SK, Viswanadham N, Al-Muhtaseb AH. Enhanced production of high octane gasoline blending stock from methanol with improved catalyst life on nano-crystalline ZSM-5 catalyst. *Journal of Industrial and Engineering Chemistry* 2014;**20**: 2876–82. doi:10.1016/j. jiec.2013.11.021

[66] Viswanadham N, Saxena SK, Garg MO. Octane number enhancement studies of naphtha over noble metal loaded zeolite catalysts. *Journal of Industrial and Engineering Chemistry* 2013;**19**: 950–5. doi:10.1016/j.jiec.2012.11.014

[67] Kish SS, Rashidi A, Aghabozorg HR, *et al*. Increasing the octane number of gasoline using functionalized carbon nanotubes. *Applied Surface Science* 2010;**256**: 3472–7. doi:10.1016/J. APSUSC.2009.12.056

[68] Hosseini SH, Taghizadeh-Alisaraei A, Ghobadian B, *et al*. Performance and emission characteristics of a CI engine fuelled with carbon nanotubes and diesel-biodiesel blends. *Renewable Energy* 2017;**111**: 201–13. doi:10.1016/J.RENENE.2017.04.013

6 Effects of gasoline and its additives on health and environment

6.1 Introduction

A combination of more than 250 hydrocarbons such as cycloalkanes, alkenes, and aromatic hydrocarbons including toluene, xylene, and benzene – which are mainly carcinogenic – make up gasoline. Aside from gasoline's properties, multiple additives might be blended that affect its characteristics and toxicity. Gasoline has seen a great deal of development in the last century and its current status on the market is the result of a combination of refinery procedures, engine design, and increased awareness of the health and environmental implications. To assure a high degree of product consistency in the marketplace, considering the complexity of contemporary engines, many of these designs and manufacturing innovations either arose from or were the outcome of many standards and regulations [1–3].

Gasoline and other fossil fuel consumptions are responsible for the majority of today's air pollution. Fortunately, critical causes of ambient air pollution have the potential to be effectively mitigated via the implementation of laws and investments in cleaner modes of mobility, energy-efficient housing, power generation, industry, and better management of municipal garbage. Having said that, according to the World Health Organization (WHO), about 4.2 million people worldwide lose their lives annually as a direct consequence of exposure to outdoor air pollution. Their cause of death is usually but not limited to cardiovascular diseases such as stroke and heart disease, lung cancer, and chronic respiratory illnesses. Over 90% of the world's population breathes in air that does not comply with the WHO Air Quality Guidelines [4].

Regarding climate change and global warming as consequences of greenhouse gasses that are produced mainly due to industrial, transportation, and agricultural activities, not only do they affect soil, ecosystems, biodiversity, and freshwater resources, but they can also increase the risk of infectious diseases alongside with noncommunicable diseases [5]. Climate change has both acute and chronic impacts on the environment. Immediate effects, such as extreme weather events including hurricanes, flooding, drought, and heat waves, occur due to the increase in average temperature; long-term effects lead to reduced water availability, soil drying out, changes in the amount of arable land for farming, increased pollution, and the creation of habitats that are conducive to the spread of human and animal pathogens via insect vectors [6].

https://doi.org/10.1515/9783110999969-006

6.2 Gasoline and air pollution

Toxicology testing on gasoline and its evaporative emissions, as well as refinery process streams, has yielded considerable hazard characterization data. The burning of diesel and gasoline, which produce more than 75% of all atmospheric pollutants, is a significant cause of air pollution in cities. The European Union considers primary pollutants as nitrogen oxides (NO_x), sulfur dioxide (SO_2), ammonia (NH_3), ozone, particulate matter (PM), carbon monoxide (CO), non-methane volatile organic compounds (NMVOCs), and methane (CH4). A substantial growing body of evidence shows that human exposure to vehicle pollutants causes various cancerous and noncancerous diseases, including cardiopulmonary disorders, reproductive dysfunction, neurological disorders, dementia, auditory dysfunction, and multiple malignancies, thereby increasing mortality and morbidity rates [2, 3, 7–10].

6.2.1 Gasoline's effects on health

Different gasoline exposure pathways exist. Gasoline starts vaporizing at room temperature; therefore, inhalation is a common route. In confined or poorly ventilated spaces, its fumes may induce asphyxiation. Respiratory symptoms may include pneumonitis, pulmonary irritation (irritant-induced asthma), and respiratory failure. Gasoline sniffing is thought to be happening in specific regions and cultures. More than half of the children in multiple indigenous communities in Australia, America, and Canada have reported that they had sniffed gasoline at least once. Its acute effects include euphoria, increased libido, and ataxia. In addition, gasoline's chronic exposure and abuse have been related to neurological alterations and even encephalopathy [1, 11].

Chemical burns and dermatitis are most often caused by gasoline spills and splashes, which are difficult to forecast or avoid. At petrol stations, a significant number of these injuries occur to youngsters. Gasoline can induce dermatitis and nail disorders, especially in station or exposed workers [12, 13]. Gasoline ingestion happens predominantly in children under five years old. Gasoline ingestion may lead to hydrocarbon poisoning and sometimes pneumonitis or pulmonary aspiration; 350 gr of it could be fatal for adults. Intoxication symptoms include nausea, vomiting, alterations in mental status, seizures, and circulatory failure [1, 13].

6.2.2 Non-methane volatile organic compounds

NMVOCs contain a broad spectrum of compounds with varying characteristics, but they all have one thing in common: they all are volatile and organic. Some NMVOCs are not harmful per se, but their reaction in the air could produce ozone which has devastating effects on the environment. Although it might seem that no serious health consequences are expected from typical unleaded gasoline use, exposure to benzene, toluene, xylene, butadiene, and polycyclic aromatic hydrocarbons may raise the risk of cancer formation and encourage cancer growth and metastasis [3].

6.2.3 Nitrogen oxides

Nitrogen oxides (NO_x) along with NMVOCs form the majority of air pollutants. Their principal origin is road transport vehicles in Europe and the United States. Exposure to NO_x emissions has been linked with respiratory disorders such as bronchitis and asthma in children. Acid rains develop due to the interaction of NO_x, water, and oxygen in the atmosphere, which are not only harmful to people but also hazardous to the ecosystem and environment [14].

6.2.4 Ozone

Ozone is produced when nitrogen oxides and ultraviolet (UV) light oxidate NMVOCs photochemically. Stratospheric ozone has a protective effect on the Earth; however, tropospheric (lower layer) ozone is considered an air pollutant since it may cause damage to plants via oxidation processes, hence wreaking havoc on crops and forests, as well as affecting people breathing it whom their premature death has been suggested to have a cardiovascular and pulmonary basis. Annually, ozone exposure leads to 470,000 respiratory deaths. This pollutant is unusual because it rises in the sunshine and falls on cloudy days because of its dependency on UV radiation [15–17].

6.2.5 Particulate matters

PMs are emitted directly into the air and are directly linked to ophthalmic and respiratory disorders such as asthma and premature death [15]. Children's lung function has decreased in communities with higher PM. Moreover, it has been indicated that this alteration can be reversible when children go to societies with lower PM levels [18].

Nitrate particles that are emitted from burning fuels together with black carbon (BC) (soot), ammonia, moisture, and other compounds are key components of PMs that have diameters less than 2.5 micrometers ($PM_{2.5}$). $PM_{2.5}$ has been correlated with autism, ADHD, and asthma in children [19]. BC particles are mainly created because of incomplete combustion of fossil fuels and biomass burning and cause many health-related issues in humans and animals. BC has a noticeable impact on the environment regarding global warming: it absorbs solar energy, controls cloud formation, and facilitates snow melting.

Similar to $PM_{2.5}$, components of PM_{10} (PM that has diameters less than 10 micrometers) can be both organic and inorganic. Some studies have indicated a link between multiple sclerosis and PM_{10} exposure [20].

6.2.6 Sulfur

6.2.6.1 Sulfur dioxide
Sulfur dioxide (SO_2) is a hazardous gas produced as a by-product of burning fossil fuels and other industrial processes. All living beings, including humans, animals, and plants, are influenced. People who already have pulmonary diseases, the elderly, and children are the most at risk of experiencing adverse effects, making them the most susceptible population. During the 1960s, the concentrations of air pollutants, particularly that of sulfur dioxide (SO2), were extremely high in many industrial cities; hence the prevalence of asthma and chronic bronchitis increased among their residents. Sulfur dioxide emissions in industrialized regions cause respiratory irritation, bronchitis, mucus formation, and bronchospasm. In addition, redness of the skin, cloudiness of the eyes, and a worsening of underlying cardiovascular disease have all been reported as side effects. It would seem that emissions of sulfur dioxide cause environmental damage, such as the acidity of soil and precipitation of acid [21, 22].

6.2.6.2 Sulfur hexafluoride
Sulfur hexafluoride (SF_6) is a potent greenhouse and non-combustible gas, and because of the latter, it remains stable at ultrahigh temperatures. Accordingly, due to its insulation characteristics, it is extensively used in industries and power plants [23]. SF_6 gas has almost no acute effect on the environment, and since it is a non-toxic gas, it does not usually pose a significant risk regarding health issues. Nevertheless, because of its enormous contribution to global warming – it is 23,900 times more potent than CO_2 – the long-term consequences might be dire: Harmful material buildup in food and water by reducing water availability and altering fertile soils are some examples. It will facilitate the spread of diseases between people and animals, resulting in an increase in the frequency and severity of outbreaks and alterations in transmissible illnesses [6, 24].

6.2.7 Carbon monoxide

When burned improperly, fossil fuel generates CO in the combustion process. Nausea, vomiting, weakness, and neurological symptoms such as headaches, dizziness, and loss of consciousness are some of the symptoms of CO poisoning.

CO poisoning may also cause death. Hemoglobin's affinity for CO is much higher than its affinity for oxygen. Poisoning from CO may occur if a person is exposed to significant amounts of gas for an extended period of time.

The consequences of CO are strongly connected to global warming. Temperature increases in the ground, and the water might lead to severe weather conditions or storms. Contrary to its disadvantages, its potential to increase plant growth has been stated in some studies [22].

6.2.8 Methane

Methane is mainly produced in industrial agriculture, fossil fuels, especially from natural gas, wastes, and natural resources such as wetlands. It is one of the principal greenhouse gases that has prominently contributed to global warming [25].

6.2.9 Ammonia

Ammonia (NH_3) emissions are mainly produced from agricultural activity. Its emissions have increased prominently after the twentieth century. Ammonia and ammonium (NH_4^+) high deposition has been demonstrated to cause environmental disturbances such as acidification, eutrophication, and biodiversity reduction [26]. Ammonia, combined with other emissions such as NO_x and SO_2, may produce aerosols that are regarded as $PM_{2.5}$. $PM_{2.5}$ exposure effects mainly include pulmonary disorders.

6.3 Gasoline additives

6.3.1 Organometallic compounds

6.3.1.1 Tetraethyl lead

Leaded gasoline can cause lead poisoning; however, tetraethyl lead as a significant additive has been phased out globally beginning in the 1970s. Lead can be excreted in breast milk. Low levels of lead exposure can prominently influence children's cognitive and educational abilities, resulting in neurodevelopment delay. Along with neurological disorders, memory loss, and seizures, lead poisoning contributes to decreased lung function in children. Central nervous system (CNS) disorders, cardiovascular abnormalities, renal disorders, depression, and reduced libido ensue lead poisoning in adults [27, 28].

6.3.1.2 Other organometallic compounds

Considering safer organometallic additives, up to 30 ppm of ferrocene has been identified to show no adverse carcinogenic and toxic effects. It has been indicated that ferrocene, in the form of pills, demonstrates antimalarial and anticancer characteristics [29, 30]. Methylcyclopentadienyl manganese tricarbonyl (MMT) is used in gasoline in Canada, the United States, South America, Europe, Asia, and Africa. The usage of MMT improves the environment by cutting tailpipe emissions of toxins such as benzene, formaldehyde, and acetaldehyde, as well as NO_x and greenhouse gases. Diet is the primary source of manganese exposure, although it can be inhaled in a lesser amount too. Manganese has been shown to induce neurotoxicity [31, 32].

6.3.2 Ethers

After cessation of applying TEL, methyl tertiary-butyl ether (MTBE) was started to substitute leaded gasoline. Commercialized MTBE production began in 1979 in the United States and in 1973 in Europe. When it was first brought onto the market in Italy in the late 1970s, it was quickly followed by other European nations. MTBE poses a more significant threat to groundwater quality than other gasoline additives. In contrast to benzene and toluene, MTBE pollution from spills or leaked undersea fuel reservoirs may spread over a broader region due to its water solubility. Human exposure to fuel oxygenates may occur via direct consumption, as well as through cutaneous contact and inhalation when water sources are polluted. Recent studies have shown a link between MTBE exposure and metabolic syndrome, increasing the risk of cardiac disease, stroke, and diabetes mellitus type 2. MTBE-induced obesity and chronic inflammation are associated with multiple organ tumors as well. Although the United States phased out this additive in 2006, it is still being used in Europe, Asia, and Latin America. MTBE concentration was not statistically significant in Germany, France, Sweden, Denmark, Finland, and the United Kingdom compared to the United States [33–36].

Regarding other examples of ether additives, the production of gasoline combined with ethyl tert-butyl ether (ETBE) is on the rise to minimize greenhouse gas emissions. The scientific literature on ETBE biodegradation in soils and groundwater is relatively limited. Despite the variations in their physical and chemical characteristics, it seems that ETBE behaves in groundwater in a manner similar to MTBE [37, 38]. Similar to MTBE, tert-amyl methyl ether (TAME) has toxic properties. TAME has been found in animal studies to be almost 10 times more potent than MTBE and a more powerful CNS depressant. Nonetheless, it seems that the fuel ethers represent no extra danger to the development of liver damage. Exposure to these ethers at low concentrations will not likely result in a considerable rise in the human body burden of these chemicals, making any harmful consequences of these ethers in humans improbable under actual exposure settings [39–41].

6.3.3 Biofuels

The main sources of biofuels include plants, animals, and algae that are heavily used in synthesizing bioethanol and biobutanol. Animal, plant, and algal fats and oils serve as the starting point for the generation of biofuels. The health impact assessment of first-generation liquid biofuels shows some advantages regarding carcinogenic compounds and nonionizing radiation. However, considering respirable compounds, health impacts may vary due to the feedstock origin of the biofuel. Following cellulose, the second most significant source of sugars for biofuel production is mannans from softwood. Mannan-degrading enzymes are an essential class of enzymes that can be employed for both pre-treatment and complete sugar release in the synthesis of second-generation biofuels, as well as the creation of possibly health-promoting mannooligosaccharides (MOS). Some studies have shown that MOS may have a positive impact on animal health as well as human health. Notably, alternative fuels like biodiesel and bioethanol are not necessarily advantageous in terms of human health impacts solely because they are derived from a renewable source; nonetheless, their influence on the environment is thought to be significantly better than fossil fuels [42–44].

6.3.4 Ethanol

A growing body of studies suggests that combining ethanol with gasoline decreases hazardous chemical emissions, particularly aromatics during cold start and highway driving. Because cytotoxic matters, reactive oxygen species, and carcinogenic compounds such as dehydrotropyl ions and methyl sulfate are greatly decreased owing to the application of ethanol; it may be argued that ethanol blending in gasoline is advantageous to human health and environment [3, 45, 46].

References

[1] Gasoline, Automotive | Medical Management Guidelines | Toxic Substance Portal | ATSDR. https://wwwn.cdc.gov/TSP/MMG/MMGDetails.aspx?mmgid=465&toxid=83 (accessed 30 Jun 2022).

[2] Swick D, Jaques A, Walker JC, *et al*. Gasoline toxicology: Overview of regulatory and product stewardship programs. *Regulatory Toxicology and Pharmacology* 2014;**70**:S3–12. doi:10.1016/J.YRTPH.2014.06.016

[3] Mueller S, Dennison G, Liu S. An Assessment on Ethanol-Blended Gasoline/Diesel Fuels on Cancer Risk and Mortality. *International Journal of Environmental Research and Public Health* 2021;**18**. doi:10.3390/IJERPH18136930

[4] Ambient air pollution. https://www.who.int/teams/environment-climate-change-and-health /air-quality-and-health/ambient-air-pollution (accessed 3 Jul 2022).

[5] Rockström J, Steffen W, Noone K, et al. A safe operating space for humanity. *Nature 2009* *461:7263* 2009;**461**: 472–5. doi:10.1038/461472a

[6] Rossati A. Global Warming and Its Health Impact. *The International Journal of Occupational and Environmental Medicine* 2017;**8**: 7. doi:10.15171/IJOEM.2017.963

[7] Rossner P, Cervena T, Vojtisek-Lom M, et al. The Biological Effects of Complete Gasoline Engine Emissions Exposure in a 3D Human Airway Model (MucilAirTM) and in Human Bronchial Epithelial Cells (BEAS-2B). *International Journal of Molecular Sciences* 2019;**20**. doi:10.3390/IJMS20225710

[8] Roggia SM, de França AG, Morata TC, et al. Auditory System Dysfunction in Brazilian Gasoline Station Workers. *Int J Audiol* 2019;**58**: 484. doi:10.1080/14992027.2019.1597286

[9] Sources and emissions of air pollutants in Europe – European Environment Agency. https://www.eea.europa.eu/publications/air-quality-in-europe-2021/sources-and-emissions-of-air (accessed 3 Jul 2022).

[10] Peters R, Ee N, Peters J, et al. Air Pollution and Dementia: A Systematic Review. *Journal of Alzheimer's Disease* 2019;**70**:S145. doi:10.3233/JAD-180631

[11] Cairney S, Maruff P, Burns C, et al. The neurobehavioural consequences of petrol (gasoline) sniffing. *Neuroscience and Biobehavioral Reviews* 2002;**26**: 81–9. doi:10.1016/S0149-7634 (01)00040-9

[12] Jia X, Xiao P, Jin X, et al. Adverse effects of gasoline on the skin of exposed workers. *Contact Dermatitis* 2002;**46**: 44–7. doi:10.1034/J.1600-0536.2002.460109.X

[13] Drago DA. Gasoline-related injuries and fatalities in the United States, 1995–2014. *International Journal of Injury Control and Safety Promotion* 2018;**25**: 393–400. doi:10.1080/17457300.2018.1431947

[14] Syed S, Renganathan M. NOx emission control strategies in hydrogen fuelled automobile engines. *Australian Journal of Mechanical Engineering* 2022;**20**: 88–110. doi:10.1080/14484846.2019.1668214

[15] Koolen CD, Rothenberg G. Air Pollution in Europe. *Chemsuschem* 2019;**12**: 164. doi:10.1002/CSSC.201802292

[16] Madaniyazi L, Nagashima T, Guo Y, et al. Projecting ozone-related mortality in East China. *Environment International* 2016;**92–93**: 165–72. doi:10.1016/J.ENVINT.2016.03.040

[17] Silva RA, West JJ, Zhang Y, et al. Global premature mortality due to anthropogenic outdoor air pollution and the contribution of past climate change. *Environmental Research Letters* 2013;**8**: 034005. doi:10.1088/1748-9326/8/3/034005

[18] Perera FP. Multiple threats to child health from fossil fuel combustion: Impacts of air pollution and climate change. *Environmental Health Perspectives* 2017;**125**: 141–8. doi:10.1289/EHP299

[19] Perera F, Ashrafi A, Kinney P, et al. Towards a fuller assessment of benefits to children's health of reducing air pollution and mitigating climate change due to fossil fuel combustion. *Environmental Research* 2019;**172**: 55–72. doi:10.1016/J.ENVRES.2018.12.016

[20] Abbaszadeh S, Tabary M, Aryannejad A, et al. Air pollution and multiple sclerosis: a comprehensive review. *Neurological Sciences* 2021;**42**: 4063–72. doi:10.1007/S10072-021-05508-4

[21] Shima M. Health Effects of Air Pollution: A Historical Review and Present Status. *Nihon Eiseigaku Zasshi* 2017;**72**: 159–65. doi:10.1265/JJH.72.159

[22] Manisalidis I, Stavroupoulou E, Stavropoulos A, et al. Environmental and Health Impacts of Air Pollution: A Review. *Frontiers in Public Health* 2020;**8**: 14. doi:10.3389/FPUBH.2020.00014/BIBTEX

[23] Yokomizu Y, Terada M, Kodama N, et al. Predominant reaction products and dielectric breakdown properties of gas mixtures consisting of SF6 and ablation products of C2F4/BN in

the temperature range of 300–3000 K. *Journal of Physics D: Applied Physics* 2021;**54**: 165204. doi:10.1088/1361-6463/ABDA80

[24] Han S-H, Seon HS, Shin P-K, *et al.* Conversion of SF 6 by thermal plasma at atmospheric pressure.

[25] Sources of methane emissions – Charts – Data & Statistics – IEA. https://www.iea.org/data-and-statistics/charts/sources-of-methane-emissions-2 (accessed 3 Jul 2022).

[26] Zhang Z, Zeng Y, Zheng N, *et al.* Fossil fuel-related emissions were the major source of NH3 pollution in urban cities of northern China in the autumn of 2017. *Environmental Pollution* 2020;**256**. doi:10.1016/J.ENVPOL.2019.113428

[27] Hon KL, Fung CK, Leung AKC. Childhood lead poisoning: An overview. *Hong Kong Medical Journal* 2017;**23**: 616–21. doi:10.12809/HKMJ176214

[28] Miracle VA. Lead Poisoning in Children and Adults. *Dimensions of Critical Care Nursing* 2017;**36**: 71–3. doi:10.1097/DCC.0000000000000227

[29] Peters L, Ernst H, Koch W, *et al.* Investigation of chronic toxic and carcinogenic effects of gasoline engine exhausts deriving from fuel without and with ferrocene additive. *Inhalation Toxicology* 2000;**12**: 63–82. doi:10.1080/08958378.2000.11463200

[30] Peter S, Aderibigbe BA. Ferrocene-Based Compounds with Antimalaria/Anticancer Activity. *Molecules* 2019;**24**. doi:10.3390/MOLECULES24193604

[31] Pfeifer GD, Roper JM, Dorman D, *et al.* Health and environmental testing of manganese exhaust products from use of methylcyclopentadienyl manganese tricarbonyl in gasoline. *Science of the Total Environment* 2004;**334–335**: 397–408. doi:10.1016/J. SCITOTENV.2004.04.043

[32] Dobson AW, Erikson KM, Aschner M. Manganese neurotoxicity. *Ann N Y Acad Sci* 2004;**1012**: 115–28. doi:10.1196/ANNALS.1306.009

[33] Achten C, Kolb A, Püttmann W, *et al.* Methyl tert-butyl ether (MTBE) in river and wastewater in Germany. 1. *Environmental Science and Technology* 2002;**36**: 3652–61. doi:10.1021/ES011492Y

[34] Chapter 11 – MTBE in petrol as a substitute for lead – European Environment Agency. https://www.eea.europa.eu/publications/environmental_issue_report_2001_22/issue-22-part-11.pdf/view (accessed 30 Jun 2022).

[35] Silva LK, Espenship MF, Pine BN, *et al.* Methyl Tertiary-Butyl Ether Exposure from Gasoline in the U.S. Population, NHANES 2001–2012. *Environmental Health Perspectives* 2019;**127**. doi:10.1289/EHP5572

[36] Rais Y, Drabovich AP. Gasoline-derived methyl tert-butyl ether as a potential obesogen linked to metabolic syndrome. *Journal of Environmental Sciences (China)* 2020;**91**: 209–11. doi:10.1016/J.JES.2020.02.008

[37] Okamoto K, Hiramatsu M, Hino T, *et al.* Evaporation characteristics of ETBE-blended gasoline. *Journal of Hazardous Materials* 2015;**287**: 151–61. doi:10.1016/J.JHAZMAT.2015.01.024

[38] Thornton SF, Nicholls HCG, Rolfe SA, *et al.* Biodegradation and fate of ethyl tert-butyl ether (ETBE) in soil and groundwater: A review. *Journal of Hazardous Materials* 2020;**391**. doi:10.1016/J.JHAZMAT.2020.122046

[39] Elovaara E, Stockmann-Juvala H, Mikkola J, *et al.* Interactive effects of methyl tertiary-butyl ether (MTBE) and tertiary-amyl methyl ether (TAME), ethanol and some drugs: Triglyceridemia, liver toxicity and induction of CYP (2E1, 2B1) and phase II enzymes in female Wistar rats. *Environmental Toxicology and Pharmacology* 2007;**23**: 64–72. doi:10.1016/J. ETAP.2006.07.003

[40] Aivalioti M, Pothoulaki D, Papoulias P, *et al.* Removal of BTEX, MTBE and TAME from aqueous solutions by adsorption onto raw and thermally treated lignite. *Journal of Hazardous Materials* 2012;**207–208**: 136–46. doi:10.1016/J.JHAZMAT.2011.04.084

[41] Dekant W, Bernauer U, Rosner E, *et al.* Toxicokinetics of ethers used as fuel oxygenates. *Toxicology Letters* 2001;**124**: 37–45. doi:10.1016/S0378-4274(00)00284-8

[42] Fink R, Medved S. Health impact assessment of liquid biofuel production. *International Journal of Environmental Health Research* 2013;**23**: 66–75. doi:10.1080/09603123.2012.699030

[43] Yamabhai M, Sak-Ubol S, Srila W, *et al.* Mannan biotechnology: From biofuels to health. *Critical Reviews in Biotechnology* 2016;**36**: 32–42. doi:10.3109/07388551.2014.923372

[44] Shahid MK, Batool A, Kashif A, *et al.* Biofuels and biorefineries: Development, application and future perspectives emphasizing the environmental and economic aspects. *Journal of Environmental Management* 2021;**297**. doi:10.1016/J.JENVMAN.2021.113268

[45] Clark NN, McKain DL, Klein T, *et al.* Quantification of gasoline-ethanol blend emissions effects. *Journal of the Air and Waste Management Association* 2021;**71**: 3–22. doi:10.1080/10962247.2020.1754964

[46] Agarwal AK, Singh AP, Gupta T, *et al.* Toxicity of exhaust particulates and gaseous emissions from gasohol (ethanol blended gasoline)-fuelled spark ignition engines. *Environmental Science: Processes and Impacts* 2020;**22**: 1540–53. doi:10.1039/D0EM00082E

7 Future perspectives and conclusion

7.1 Key concerns and other research requirements

Reducing exhaust emissions will be the main field of study for ongoing research. New formulas for blending fuels (gasoline and diesel) leading to lower soot contamination could be contemplated in future studies. The following research areas have excellent capability and capacities for future research in this regard. Energy and cost optimization use different nano-additives. Researchers should focus on nano-metallic additives, which have two critical advantages; they reduce soot emission (by splitting hydrogen of the water, achieving complete combustion) and improve engine performance BTE. It helps cost optimization because using alcohol to reduce air pollution in this situation is not essential. Researchers should pay more attention to gasoline's chemical-physical properties as they directly affect air pollution and engine performance. Many studies have been done at various speeds, engine loads, pressures, injection timings, and other laboratory conditions [1–3]. This study paves the way for considering other conditions like different engine materials or conducting more experiments with engines with various volumes.

7.2 Conclusion

This book focused on the influence of different additives on engine performance, exhaust emission, and chemical and physical properties of gasoline (precisely the octane number). The following are the main conclusions:

1. Adding alcohol reduces CO and HC emissions, which is the main reason for enhancing CO_2. It is experimentally concluded that oxygenated additives, especially alcohols with higher oxygen content such as methanol and ethanol, improve combustion efficiency. Therefore, it is one of the causes that reduce particulate matter, HC, and CO emissions. Higher latent heat vaporization of alcohols compared to gasoline and lower boiling points are the other reasons for these results. On the other hand, increasing oxygen content and reaction with CO brings about CO_2 emission. The alternative way to reduce CO_2 is by using biofuel because raw materials used for biofuel may capture CO_2 from the atmosphere.

2. NOx emission is enhanced by using alcohol as additives because of alcohol's oxygen content and some of their chemical and physical properties like latent heat of vaporization, triggering higher temperatures for making NOx. Previously, mechanical and chemical engineers tried to eliminate this problem using EGR.

https://doi.org/10.1515/9783110999969-007

3. Nano-additives are used for different purposes, such as controlling air pollution and improving engine and fuel performance. Before using nano-additives, we should analyze them and study their influence on the environment because some of them, such as silver oxides, might decline soot emissions. On the contrary, these types of nano-additives harm other living beings, especially sea creatures.

4. The octane number of gasoline would be improved by alcohol. Using methanol as an additive can improve the octane number. Nevertheless, blending ethanol with gasoline is more harmful to the ozone layer, making climatic changes. Moreover, alcohols influence other gasoline properties like density, viscosity, and latent heat of vaporization due to their properties' differences from gasoline.

5. Other additives such as MTBE, DIPE, and ETBE may improve the octane number and BTE, but they might harm the environment by increasing soot emission.

References

[1]   Örs I, Sarıkoç S, Atabani AE, *et al.* The effects on performance, combustion and emission characteristics of DICI engine fuelled with TiO2 nanoparticles addition in diesel/biodiesel/n-butanol blends. *Fuel* 2018;**234**: 177–88. doi:10.1016/j.fuel.2018.07.024

[2]   Sarıkoç S, Örs İ, Ünalan S. An experimental study on energy-exergy analysis and sustainability index in a diesel engine with direct injection diesel-biodiesel-butanol fuel blends. *Fuel* 2020;**268**. doi:10.1016/j.fuel.2020.117321

[3]   Doğan B, Erol D, Yaman H, *et al.* The effect of ethanol-gasoline blends on performance and exhaust emissions of a spark ignition engine through exergy analysis. *Applied Thermal Engineering* 2017;**120**: 433–43. doi:10.1016/j.applthermaleng.2017.04.012

Index

https://doi.org/10.1515/9783110999969-008